THE EVOLUTION OF
DEVELOPMENT

THE EVOLUTION OF DEVELOPMENT

THREE SPECIAL LECTURES GIVEN AT UNIVERSITY COLLEGE, LONDON

BY

JOHN TYLER BONNER, Ph.D.

Associate Professor of Biology, Princeton University

CAMBRIDGE

AT THE UNIVERSITY PRESS

1958

PUBLISHED BY
THE SYNDICS OF THE CAMBRIDGE UNIVERSITY PRESS
Bentley House, 200 Euston Road, London, N.W. 1
American Branch: 32 East 57th Street, New York 22, N.Y.

©

CAMBRIDGE UNIVERSITY PRESS
1958
except in the U.S.A.

Printed in Great Britain at the University Press, Cambridge
(Brooke Crutchley, University Printer)

CONTENTS

ACKNOWLEDGMENTS   *page* vii

LECTURE I   THE ORIGIN OF DEVELOPMENT   I

LECTURE II   THE FUNCTION OF DEVELOPMENT   34

LECTURE III   THE EXTENSION OF DEVELOPMENT   69

INDEX   101

ACKNOWLEDGMENTS

The content of these lectures has been slowly forming for some time and as a consequence I hardly have a friend to whom I am not indebted in some way. There are certain people who at some stage in the progress were especially generous with their time and I should like to thank them particularly. Professor T. M. Sonneborn and an anonymous reader were largely responsible for giving me some perspective in my early fumblings, and Professors E. G. Butler and D. R. Stadler spent many hours (to my great profit) with the manuscript at a later stage. Furthermore, Professor D. R. Griffin was most helpful in his comments on the section concerning animal behaviour.

It is a special pleasure to thank Professor G. P. Wells and the other members of the University of London who were so kind and hospitable to me during the period these lectures were delivered.

I am grateful to the following individuals for permission to use their illustrations: P. Brien, E. Fauré-Fremiet, C. B. Metz, J. R. Raper, K. B. Raper, V. Tartar, N. Tinbergen, C. H. Waddington, and P. B. Weisz. Also, I gratefully acknowledge the following journals and publishing houses for permission to use the illustrations listed: Longmans, Green and Co., Figs. 6, 15; Oxford University Press, Figs. 28–33; *Quarterly Review of Biology*, Figs. 18–20, 23; *Scientific American*, Fig. 16. And I wish to thank Marcia J. Shaw for her fine original drawings: Figs. 2, 5, 7, 8, 12, 17, 18, 19, 20, 23, 25.

Finally I am indebted to the Eugene Higgins Fund of Princeton University for financial assistance in the preparation of this book.

J. T. B.

PRINCETON, NEW JERSEY
April 1957

vii

THE ORIGIN OF DEVELOPMENT

I DO not propose to say anything new or original in these lectures. But I am a great believer in saying familiar, well-known things backwards and inside out, hoping that from some new vantage point the old facts will take on a deeper significance. It is like holding an abstract painting upside down; I do not say that the meaning of the picture will suddenly be clear, but some of the structure of the composition that was hidden may show itself.

The abstraction I hold before you is the development of living organisms. The novel view that we shall take of this well-worn and battered subject is its evolution. There has been, after all, an evolution of development along with the evolution of living organisms, and perhaps if we look at this aspect of development we shall see some of the problems of mechanism more clearly. There is no denying that development remains one of the largest collections of unsolved problems in biology today. It has yet to be touched with a flash of light in the way that the gene theory has illuminated the mechanism of heredity.

One of the limitations, as Huxley[1] has recently pointed out, is that we tend to think of development solely in terms of immediate causes, as the names 'causal embryology' or 'epigenetics' imply, and neglect the fact that development is adaptive and that there is a 'purpose', a 'goal' which is the final form and function of the adult. The criticism is that we should take a more distant and encompassing view, and such is the intent of these brief and superficial lectures.

This first lecture is on the origin of development. To put the matter quite simply, many small unicellular organisms have little that can be construed as development, for the adult and

[1] *Nature*, **177**, 807–9 (1956).

I

the embryo are approximately the same except for a difference in size. However, with an over-all increase in size of the adult, as must have occurred during the course of evolution, there has also arisen a development from germ to maturity. The reasons that this is so are undoubtedly to be found in the examination of the effects of natural selection and we shall first dwell on this aspect of the 'why' of development. Then, as one might expect, development has not originated once in a common ancestor of all larger organisms, but a number of times, and I shall describe some of the different kinds of development that have been independently devised during the course of early evolution.

The second lecture will go directly to the core of the problem: in what way does this evolutionary approach help us to understand the mechanisms of development? It gives a perspective and shows the common function of all the different kinds of development. It illuminates by underscoring the variety of ways in which living substance can mechanically achieve a single purpose. This approach is especially effective if again one imagines how such mechanisms could have arisen.

In the examination of the common function of all developments there emerges the fact that one of the prime essentials in the orderly spacing and patterning is a communication between parts. In the last lecture I shall show that such communication is not confined to the development of one individual, but to the association of separate individuals as well. In either case there is a common selective advantage of efficiency by unification and it means that in a broad sense the principles of individual development apply also to the association between organisms.

This then is a further insight into the phenomenon of development that may be gained from a consideration of evolution.

To begin I should like to remind you of a fact of which you are already quite well aware (unless you are an anti-Darwinian, which is unlikely these days). It is rather like reminding you that you have to breathe in order to obtain enough oxygen, for, common as the fact may be, we do not think of it each time we take a breath. Nor is it likely that each time you look at a biological structure or activity you say to yourself that it has arisen by natural selection. But the fact is that selection is a supreme and all-important principle which has channelled and governed every aspect of life that exists today. It must have operated from the very beginning of organic evolution, but if we push back even further to the chemical origins of life then we lose sight of selection as a guide to progress. Let us examine that early transition in more detail, for it helps to emphasize the way in which selection operates.

Consider a chemical solution, for example, sea water, which has a variety of different substances dissolved in it. The chemicals may be capable of appearing or disappearing; they may appear by chemical synthesis and disappear by a chemical degradation. But there is no conceivable way in which selection could be operating in these changes. They can only be accurately described in terms of chemical equilibria, that is to say, the changes are dependent on the concentration of the reactants and the speed of the reaction, which in turn is dependent upon the temperature and other conditions.

Selection involves two prime factors which this chemical system lacks: reproduction or duplication of the chemical particles, and variation among the particles in order to arrive at some basis of selection. Take, for example, sodium chloride. So far as is known one sodium chloride molecule is the same as the next, and there is no way in which this substance can synthesize itself. If one were to consider the variation as being represented by the different kinds of molecules in sea water, we are still faced with the impossibility of reproduction, and if the 'successful' molecules cannot make more of themselves, the word 'success' is empty and meaningless. Success in selection can only be measured in terms of the ability to produce more offspring: the survival of the fittest is basically a measure of the relative rates of reproducing new individuals.

The problem then of the origin of life from chemical systems involves both variation and reproduction. I do not intend here to discuss the various hypotheses of the origin of life,[1] but simply to point out that once one has achieved a molecule of the size and complexity of a protein, then variation on a large scale is already possible. There are so many different amino-acid groups in a

---

[1] See H. F. Blum (*Time's Arrow and Evolution*, 2nd ed., Princeton University Press, 1955) for a good discussion and review of this subject.

I-2

protein that by slight variation of their position and order within the molecule the number of combinations rapidly becomes quite staggering.

The reproduction of such primitive systems, however, is a more subtle matter.[1] From modern research on viruses and immunochemistry there is evidence that proteins and nucleic acids can be duplicated in all their specific detail, so that any minor variations of molecular structure are repeated in the off-spring. This duplication process, of course, needs energy, and I refer to Blum for a discussion of the various types of energy possible during the first phases of the evolution of life. In all cases of duplication that are known today (although admittedly not all are known) the immediate energy is supplied by a specially constructed machine, that is the cell. The cell is a unit of metabolism, an intricately devised motor which supplies energy for any duplication process, as well as all the other activities associated with life, including the synthesis of the many other protoplasmic constituents.

I have, in an easy sentence or two, brushed over what probably was the most important step in the whole process of evolution, the step from simple chemical compounds to a cell. This is done with a full awareness of the magnitude of the problem, but in discussing the origin of development we are concerned with that phase of evolution in which cellular organisms already exist. To place the beginning of development in its proper setting it is first necessary to consider the nature of primitive cellular reproduction and the origin of mechanisms for producing and transmitting variation.

The reproduction or duplication of cells is, at least on a descriptive level, a straightforward matter. For the moment let us neglect the nucleus, and it is well known that enucleated cells continue to metabolize. However, the respiration will not keep on indefinitely, for there are essential factors for mainten-ance and synthesis of new cytoplasmic constituents lodged in the nucleus. But the cells may divide, as was shown in enucleated sea-urchin eggs by E. B. Harvey;[2] here there is such a reserve

[1] H. F. Blum, Fifteenth Growth Symposium (1957).
[2] For references, see *The American Arbacia*. Princeton University Press (1956).

of spare parts spread out all about the cytoplasm that upon cleavage each new cell contains all the elements it needs for metabolism. If the complexity of the cell were compared to an electric calculating machine, it is not like sawing such a machine in half, which would certainly stop its effective operation, but it is more like cutting a room in half which contains hundreds of calculators, and those in each half-room will continue to operate. After division the different kinds of particles and the important structural components of the cell will be in each daughter cell. Cells always come from cells, and in the line of continuity many fine details may be directly passed on.

Now if one includes in this picture the power of synthesis, which is the normal situation with the nucleus present, then the whole process is continuous and self-propelling. The essential proteins and other complex molecules are constantly being manufactured; they are dispersed in sufficient numbers throughout the cell so that when the cell splits each part has all the elements which can be maintained or even reconstructed so as to resemble its parent.

Again in the case of single cells, we might assume that variation is at first the natural by-product of reproduction. That is to say, in the beginning the machine is so imperfect that by its very imperfection it produces variable offspring. The only difficulty is that some of this variation produced in the offspring will be lost as the offspring in turn reproduces, and so any selection that might have operated in that one generation will be futile. Selection then will tend to encourage and keep those kinds of favourable variations which are relatively permanent in the offspring, i.e. those that are inherited.

In some organisms the inheritance mechanism may be cytoplasmic. There may be elements in the cytoplasm, either particular enzymes or groups of them, or even some visible structure which can be altered or disappear, and this change is directly transmitted to the offspring through cell division. To give some well-known examples, there is for instance the loss of ability to synthesize a series of respiratory enzymes in yeast discovered by Ephrussi.[1] These 'petite' yeast cells lack a whole battery of enzymes associated with the cytochrome system.

[1] *Nucleo-cytoplasmic Relations in Micro-organisms.* Oxford (1953).

5

There is reason to suspect that certain cytoplasmic particles are lacking and that once they have disappeared the means of restoring them is also gone, so the change becomes a permanent, inherited one.

Another example might be the kappa particles of *Paramecium* elucidated by Sonneborn,[1] which are now visible and known to be involved with the killer characteristic. That is, individuals with many of these particles are killers in that they can destroy sensitive *Paramecia*, which are deficient in the particles, or which carry a different mutant form of kappa particle. It has been shown that these particles multiply at a rate which may be different from the rate of cell division, so that, by growing the organism at different rates (either by control of food or temperature), the number of particles per cell can be increased or decreased. The passage of these particles to offspring either in a sexual division or in sexual conjugation is entirely cytoplasmic. As with Ephrussi's 'petites', if there is a complete absence of kappa particles in a cell, then the cell is incapable of synthesizing them *de novo*.

One might ask why all inheritance mechanisms are not of this kind, rather than the nuclear, gene-controlled mechanism. Confining ourselves at the moment to lower organisms—protozoa, bacteria or yeasts—what special use to these organisms are nuclear genes? How did they arise by selection?

One answer might be that they arose because they were always there: they were the large molecules which were capable of primitive variation and which lived symbiotically with the energy machine. In other words, I am saying that there is no such thing as a cell without genes, but the metabolism part and the large variable molecule part arose together as one and they are inseparable. The nuclear genes are always ultimately responsible, among other things, for the synthesis of new cytoplasmic constituents (without specifying the mechanism), and the cytoplasmic parts are always responsible for giving the genes enough energy to operate. The two from the beginning have been inseparable and mutually dependent. In this view, then, cytoplasmic inheritance would merely be a secondary and

[1] 'Beyond the gene—two years later'. From Baitsell's *Science in Progress*, 7th series, pp. 167–203 (1951).

relatively unimportant mechanism that arose as the consequence of the elaboration of the cell with its many diverse, particulate constituent parts.

Another notion might be that, in the beginning of the formation of cells and primitive organisms, the large molecules capable of variation were those nucleic acids, enzymes and other proteins which are part of the metabolism machine itself. I am suggesting here, in contrast to the previous idea, that in early living forms there were no nuclear genes as fixed cell constituents, but that variability was expressed haphazardly in any of the large molecular constituents of the cytoplasm. Here then all heredity would at first have been cytoplasmic in the very way demonstrated by Ephrussi's 'petites' or Sonneborn's 'killers', and there would have been no special, restricted nuclear genes.

In the first hypothesis it is suggested that the true genes arose in the nucleus, and in the second it is suggested that large molecules in the cytoplasm gradually took on the function of genes and then became concentrated in the nucleus. There is, of course, no way of choosing between these hypotheses, but in both cases a package of nuclear genes is the end result, and now one might ask how this could have arisen by selection. The nuclear gene mechanism represents a mechanism of centralized and orderly arranged information. In one concentrated spot there is a group of molecules or a molecular aggregate which both represents the different cell constituents and can initiate the synthesis of the cell constituents, provided the proper conditions exist. So one advantage of the nuclear gene mechanism is that it is a centralized control bureau, and I think it is hardly necessary to emphasize that this would be an advantage as far as selection is concerned, for the co-ordination of many actions can best be operated through one centre. I do not mean to imply, however, that metabolism is run by the genes, because except for synthesis it is not. The genes confine their activities to the manufacture of essential parts to keep the motor going; hence they are concerned with intracellular reproduction.

Another more specific (and perhaps more significant) advantage of the nuclear apparatus is that it ensures a higher probability of equal distribution of gene material at cell division.

Not only is there equal division, but this is effected with the minimum of material; it is a precise and economical method of separating the genes. Only one of each is needed instead of a multitude which would be required in any system of random distribution.[1]

If evolution, then, is to operate in dividing single cells there must be a stable inheritance mechanism of variants and for this the nuclear gene mechanism is most suited and far outstrips cytoplasmic inheritance in importance. Furthermore, there must be some method of controlling the degree of variation; for, as Darlington[2] points out, too great a degree of variation would break up advantageous combinations, and too low a degree would never achieve the most advantageous combinations.

Sexuality is, of course, the principal means of governing the amount of variation. There are a number of interesting alternatives to sex discovered recently in viruses, bacteria and fungi, but for the present purpose a brief mention of sexual reproduction is entirely adequate. Let me remind you that there are haploid gametes which fuse, thereby uniting the genetic constitution of two different individuals. Variation is possible through mutation and recombination and both of these processes are complex and controlled. The keystone to the selective advantages of sexuality in variation production is meiosis. For here is not only the diploid number of chromosomes reduced back to the haploid (a necessary antidote to fertilization) but also the major reshuffling of traits occurs at this moment.

I find it helpful, in thinking of the function and origin of sexuality, to look at the hypothetical scheme of Beadle and Coonradt[3] for the evolution of sex in the fungi. It is of course most unlikely that this represents the origin of sexuality for other forms, but it serves as a good illustration of how control could be exerted in each step towards an increasingly efficient system of sexuality.

The scheme is illustrated in Fig. 1. In the beginning one has haploid organisms which are self-sufficient and capable of synthesizing all the necessary nutrients. By mutation some of

---

[1] I am indebted to T. M. Sonneborn for this point.
[2] *The Evolution of Genetic Systems.* Cambridge University Press (1939).
[3] *Genetics,* **24**, 291–308 (1944).

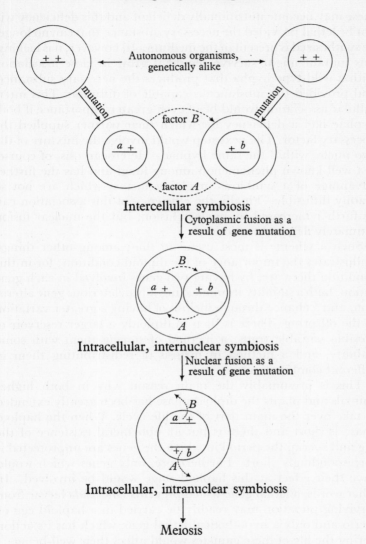

Fig. 1. Steps in the evolution of sexuality according to the
hypothesis of Beadle and Coonradt.

these may become nutritionally deficient and this deficiency will not be lethal provided the necessary substance that can no longer be synthesized is present in the medium. If, however, it is lacking, this mutant might survive should it grow in close association with a wild-type hypha that produces the necessary substance, and provided the substance is capable of diffusion. This intercellular association would be of even greater importance if both hyphae had a deficiency for which their partner supplied the necessary factor. The next step would involve the mixture of the two nuclei within the same hyphae. Heterocaryosis, of course, is a well-known phenomenon among fungi and has the further advantage of a symbiosis involving factors which are not so readily diffusible. Finally the efficiency of this association can be further increased by nuclear fusion, but the nuclear fusion ultimately demands meiosis.

Such a scheme is most useful in that, among other things, it illustrates the importance of the diploid condition; for in this condition there are, by having two genes involved in each gene action, both a stability from masking any deleterious gene alteration, and a chance through meiosis of having a greater variation in the offspring. There is then in diploidy a larger reservoir of possible variable traits, a method of storing them with some stability, and a method in meiosis of redistributing them in different combinations.

This is presumably the main reason why in both higher animals and plants the diploid phase has been greatly extended to take over the main part of the life cycle. When the haploid phase is short and there is but an ephemeral existence of the egg and sperm, the period in which the genes are unprotected is correspondingly short. Furthermore, only genes which would have their effect at this haploid stage would be involved. In other words, a lethal gene which prevents *Drosophila* larvae from surviving pupation may readily be carried in a haploid egg or sperm and only a hypothetical lethal gene which has its action during the life of these gametes would affect their well-being.

If, as in many algae and bryophytes, there is an alternation of generations, that is, a prolonged haploid phase, as well as a diploid phase, we clearly see that although the system may be less efficient in some respects to the relatively complete diploidization of higher forms, it still is such that selection may easily operate and produce successful evolutionary progress. Although the diploid state may have properties which we can rationally see have certain advantages, the fact remains that many complex organisms, such as some large algae or the males of hymeno-pterous insects, exist and develop as haploids and furthermore are often members of groups that show considerable evolutionary progress. In such alternating haploids there is still meiosis with all its advantages of recombination, and all that is lost is the buf-fering or stabilizing afforded by the exaggerated diploid phase.[1]

In the majority of organisms the sexual process concerns single cells. The gametes are single and two fuse to form one zygote. This kind of mechanism, which is in itself most complex when one considers all the details of the nuclear and chromosomal behaviour, must have had a long, slow evolutionary history. It apparently is efficient judging from its ubiquity in most animals and plants, and furthermore its very complexity would argue against radical changes appearing, which would further account for the uniformity of the nuclear mechanisms in all organisms. But the important point for our argument here is that the processes of fertilization and meiosis occur in unitary nuclei.

This simple fact—that the mechanism which produces and transmits variation uses single cells—is vital to the origin of development. Because the increase in size has, under certain conditions, adaptive value, there has been a repeated and per-sistent trend towards a multicellular condition.[2] If the organism is to be multicellular and its variation system unicellular, the direct consequence will be a development. To put the matter succinctly, development is the result of sex and size.

Like any sweeping generalization there are exceptions and some of these exceptions are in themselves of interest. First of all there are organisms, such as rotifer worms, which are minute yet possess a development. The case can be accounted for by making

[1] It may be that such haploidy is useful in purging the organism of deleterious recessives (see J. S. Huxley, *Evolution, The Modern Synthesis.* George Allen and Unwin, London (1942), p. 129). For a general discussion of haploidy and diploidy see G. L. Stebbins, Jr., *Variation and Evolution in Plants.* Columbia University Press (1950).

[2] Multinucleate coenocytes are meant to be included here. Perhaps Sach's 'multi-energid' would be more appropriate than 'multicellular'.

the easy assumption that the ancestors of the present rotifers were large and through some special ecological condition there has been a reversal by selection of the trend in size.

Another exception would be the case of fungi where there may be a joining of many nuclei in the fusion of two or more hyphae, but, as we have just mentioned, these nuclei would not fuse but remain in heterocaryon. Fusion (followed by meiosis) would only take place before ascospore formation, or as Pontecorvo and his collaborators[1] have shown there may be a diploidization in the vegetative hyphae with a subsequent return to the haploid condition by mitotic reduction. In any event there is not here a fusion of two uninucleate gametes but a mass fusion that may even involve more than two parents. Such large fusions of genetically diverse nuclei have advantages, but the heterocaryon in itself cannot recombine. Recombination is limited to the sexual and the parasexual methods mentioned above and these are always restricted to a single nucleus.

Omitting the recombination methods, a heterocaryon has a variability mechanism by blending numerous genotypes and then, by the production of large numbers of uninucleate microconidia, it can segregate these different nuclei so that in the next generation they can come together in other diverse combinations. This in itself is a non-sexual method of handling variation, but note that it also depends upon a small single-cell stage, the spore. Non-sexual methods of reproduction as well as sexual ones need a small reproductive unit and therefore also need a development. It should be added that the spores are not only small to favour segregation, but also as an adaptation for more effective dispersal. Although less euphonious than 'sex and size' it would be more appropriate to say that development is the product of 'reproduction and size'.

There is one corollary to the problem of the evolutionary origin of development and that is the origin of death. Development, in its most inclusive sense, implies not only the journey from germ to adult, but on through to senility and death. The selective advantage of death, which Weismann[2] understood so

[1] *Ann. Rev. Microbiol.* **10**, 393–400 (1956).
[2] *Essays upon Heredity and Kindred Biological Problems*, 2nd ed., chap. 1. Oxford (1891).

well, is obvious. In evolution progress is dependent on new forms and in order to clear a space for the new ones the old ones must be eliminated. This means that the generation time and the time during one generation in which the organism is capable of reproduction, are factors modified by selection. Retention of small units in the reproductive phase, increase in size of the somatic individual, development, and death, are all part of one scheme, well defined within the rules of natural selection.

The stage is now set to consider actual cases of origins of development. Each time that the multicellular condition was achieved (and it is quite reasonable to assume that this occurred repeatedly) there was an independent and new origin of development. Since this happened in early earth history, and since it left no palaeontological record, exactly how many times this happened and exactly how it happened is of course entirely a matter of speculation. All one can do is 'reconstruct the crime' from the present lower colonial forms; this excludes all the less successful attempts at colonization that have not survived the millions of years since their invention. Furthermore no effort will be made to include all the known cases that exist today, but only a few interesting and divergent ones.

At the moment I will confine the discussion to straightforward descriptions of life cycles, but in these descriptions it will be helpful to keep in mind three major constructive processes of development: growth, morphogenetic movement and differentiation. Growth is used here in the sense of an increase in living matter; it involves the intake of energy and the storing of some of that energy by the synthesis of new protoplasm. Morphogenetic movement is the migration of protoplasm which gives rise to changes in form. There is no synthesis of living material, but merely its movement from one region to another. Differentiation is the specialization of parts of the living protoplasm so that their chemical composition and structure is discrete and different. Differentiation is a manifestation of division of labour and the structures that are associated with a particular function are said to be differentiated.

These three processes may be present in unicellular organisms as well as in multicellular ones. An amoeba grows and its protoplasm moves, thereby affecting its form; the eye-spot or the flagellum of *Euglena* would be a good example of single-celled differentiation. But in these lectures we are considering all three with respect to development, and by development we mean that period in the life history of an organism where growth, mophogenetic movement and differentiation contribute towards elaborating a single-celled or small germ into a large, multicellular individual.[1]

I have attempted, in Fig. 2, to show in one scheme some of the different independent attempts at multicellularity. There is a weak effort to arrange the organisms with some phylogenetic rationale but this is not too successful, for relationships between primitive organisms is a branch of biology that might be called one of total speculation. In general outline we assume that the cells of bacteria are the most primitive type and that they gave rise, possibly through some spirochaete-like intermediate, to the basic flagellate cell. The flagellate may be assumed to be the ancestral form of all higher multicellular organisms.

In Fig. 2 the centrifugal direction represents an increase in size and number of cells or 'energids' in the somatic individuals (i.e. an increase in the amount of development). Some organisms stick closely to the flagellated cell as the primary building-block. For example (reading Fig. 2 from the top and going clockwise), there are such green algae as *Hydrodictyon*, *Pediastrum* and all the Volvocales.[2] Among non-photosynthetic forms there are the sponges, higher animals in general, different slime organisms (Myxomycetes, Plasmodiophorales, Labyrinthulales), Foraminifera, Radiolaria and various others. In many of these forms during some stages the flagella are lost and certain of the cells may become amoeboid, but the close affinity between pseudopod and flagellum is well recognized. Certain groups have no known flagellated cells in their life histories, such as the

[1] This limited definition of 'development' is used here for the specific purpose of presenting the ideas of these lectures. I should stress again that 'multicellular' is also being used in the special sense of 'multi-energid'. Perhaps even 'multi-nucleate' or 'multi-genome' would be appropriate and therefore this is a discussion of 'multi-energid development'.

[2] Some of these organisms will be described in some detail presently, while the others are merely listed here for the purpose of rounding out the chart shown in Fig. 2.

large multinucleate amoebae (e.g. *Chaos* or *Pelomyxa*) and the cellular slime moulds or Acrasiales.

The system of classification used here is on the basis of the kind of cell building-block. Another obvious close relative of flagellates are ciliates and these, by virtue of their high degree of polyploidy, qualify as multi-energids. Flagellates also were

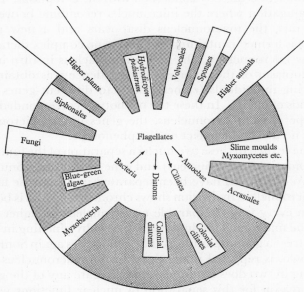

Fig. 2. A scheme representing some of the principal independent attempts at multicellularity and increase in size among lower forms. The centre circle contains one-cell forms, the shaded ring is a rough approximation of colonial forms, and the true multicellular forms extend beyond the ring.

undoubtedly ancestral to the diatoms which form peculiar and interesting colonies, but we will not have the space to include them in this discussion. The most successful, hard-walled building-block that has arisen from the flagellate cell is the filament and the filament is the basis of construction of the fungi, the coenocytic algae and the higher plants in general

In ciliates, which we shall use for our first example, there is an asexual division as well as a sexual conjugation. By dividing, the cell goes through the direct process of making two out of one, for in each daughter cell there are all the necessary genes and cytoplasmic factors. In these forms there is the interesting complication of a micro- and macronucleus. It has now been definitely demonstrated in *Paramecium aurelia* by some remarkable experiments of Sonneborn[1] that the micronucleus is entirely confined to the activities of sex and recombination while the macronucleus is directly concerned with the cytoplasm. Following conjugation where the micronuclei recombine between the conjugants, the macronucleus disappears and a new macronucleus is formed containing the new gene complex dictated by the new micronucleus. The new macronucleus in turn informs the cytoplasm of the changes. The evidence was obtained by producing individuals (clones) that had different genes in the two kinds of nuclei. In these the phenotype corresponded to the genotype of the macronucleus, the genes in the micronucleus having no detectable effect on the phenotype.

We have then in these two nuclei a separation of the function of recombination (micronucleus) and gene action (macronucleus). The two operations have been separated in a stepwise fashion. The micronucleus in division shows chromosomes and is believed to be in every way comparable to the nucleus of higher forms, while the macronucleus duplicates merely by pinching in two in the centre—a genuine amitosis. The genome, as again Sonneborn has shown, is repeated many times in the macronucleus and a pinching in two does not appear to segregate any of the genes.

The reason for this separation of nuclear functions we may assume is connected with size (although it may possibly be merely a matter of complexity). The cytoplasm has achieved such a magnitude and complexity in ciliate evolution that in order to keep the gene-cytoplasm ratio sufficiently high for the normal functioning of the animal there has been a kind of compounding of the genome in the macronucleus. But in becoming large the macronucleus has lost the ability to undergo meiosis and the result is a separation into germinal and somatic nuclei.

As far as development is concerned, unlike true multicellular forms where the nuclear material divides and keeps pace with cell division, there is here a separation of the nuclear develop-

[1] *Adv. in Gen.* **1**, 263–358 (1947). For a general discussion see G. H. Beale, *The Genetics of* Paramecium aurelia. Cambridge University Press (1954).

ment and cytoplasmic development. A micronucleus develops, following conjugation,[1] into a large macronucleus; the cytoplasm never undergoes any major change such as a reduction to small simplified gametes, but it always retains its major structure and undergoes relatively minor changes following division. It would appear that the cytoplasm has become so complex that it is incapable of a complete development, but now must be handed down, in large part directly by a massive cytoplasmic inheritance. This then would be the reason for the appearance of conjugation (where the cytoplasm remains unaltered) in the place of the usual gametes, fertilization, zygote (where there is cytoplasmic as well as gene development).[2] Therefore the significant developmental novelties that have appeared in the evolution of the ciliates are the formation of the macronucleus accompanied by such an extensive elaboration of the cytoplasm that conjugation has replaced gamete production, because the details of the cytoplasm must, in large measure, be carried directly from one generation to the next by cytoplasmic inheritance. Since there is no total cytoplasmic development, but only a nuclear one, the only structure that need remain small for sexuality and recombination is the micronucleus.

However, one must not forget a property of this elaborate cytoplasm, which has already been stressed, and that is its dependence upon the macronucleus. It is perhaps helpful to think of development as taking place in two phases: first the formation of a large macronucleus, and second that nucleus exerting its effect on the cytoplasm. The first phase occurs once after each conjugation (or autogamy); the second can occur repeatedly in successive binary fissions. Now let us examine the extent and the complexity of the cytoplasmic changes during cleavage, remembering that they are simultaneously carried on by direct cyto-

---

[1] It should be mentioned that besides conjugation there are other causes of macronuclear reorganization. The most important is autogamy or self-fertilization which will cause a heterozygous individual to become homozygous in one step. For a discussion of the role of autogamy in the life history of ciliates, see T. M. Sonneborn, *Nature*, **175**, 1100 (1955); *J. Protozool.* **1**, 38–53 (1954).

[2] A macronucleus is found in the euciliates only, for in the less complex protociliates there are two or more similar, small nuclei, gamete production and zygote formation.

plasmic inheritance as well as promoted or sustained by the genes of the macronucleus. Two cases will be given as illustrations.

The first is the hypotrich *Euplotes* (Fig. 3). On the dorsal side of *Euplotes* there are vertical rows of small non-motile bristles and these are all connected by a silver-staining network. The ventral side also possesses the network, but the bristles are lacking and instead there are a number of large compound cilia or cirri. Another type of compound cilia, the flat oral membranelles, lie about the gullet which is a groove down the ventral side. Upon the division the large C-shaped macronucleus pinches in two and the surface structures on both sides independently undergo multiplication. The dorsal bristles multiply rapidly in the region of the equator, so that by the time of division both daughter cells have a good complement of them. However, there is evidence for further multiplication during subsequent growth and also during the growth period there are appropriate expansions on the dorsal surface so that the bristles again become evenly spaced. (The expansion occurs in different regions for the anterior and posterior daughter cell, as is shown in Fig. 3.)

On the ventral surface there are two major events. The anterior daughter cell retains the old mouth, but a new mouth is developed in the posterior region and all the elaborate membranelles are formed *de novo*. Two sets of cirri anlagen develop in the middle region and as bipartition proceeds they expand. The old cirri are reabsorbed and the new ones eventually reach their proper positions in each daughter cell. Along with this expansion of the cirri anlagen there is a new silver-staining network that expands and eventually covers the ventral surface.[1]

It is my purpose, in giving these details, to show the high degree of complexity that has been achieved with the macronucleus-conjugation mechanism of ciliates. My second example is the development of a suctorian and here, in this specialized group of ciliates, we find the added feature of the formation of progeny by budding, rather than an equal division into two. The buds are essentially embryos and go through a well-defined 'larval' stage before reaching maturity.

[1] For further details, see E. Chatton and J. Séguéla, *Biol. Bull. France et Belg.* **74**, 349–442 (1940); and J. T. Bonner, *J. Morph.* **95**, 95–108 (1954).

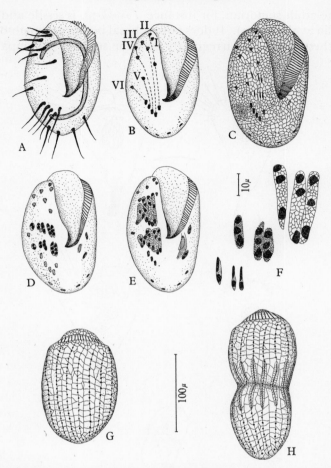

Fig. 3. The ciliate *Euplotes*. A, ventral view showing the cirri, the mouth structures and the C-shaped macronucleus; B, ventral view indicating (by dotted lines) the cirri that originate from the same group; C, silver-stained preparations showing the silver network and the beginning of the primordia of the new cirri anticipating cell division; D, E, showing the progressive stages of the cirrus development prior to division; F, three successive stages of cirri expansion (from lower left to upper right) showing the progressive widening of the spaces between the cirri; G, dorsal view showing the rows of basal bodies which lie at the base of the bristles; H, a similar view during division (note the multiplication of the basal bodies near the equatorial zone).

The adult suctorian, for instance *Tokophrya*, is sessile and has no cilia at all (Fig. 4). It does possess basal bodies or kinetosomes and these are placed irregularly over the surface. Anticipating

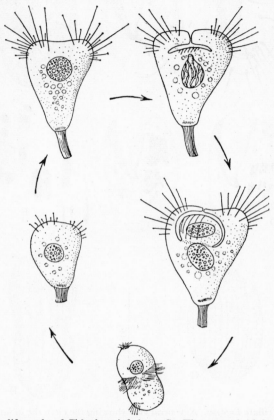

Fig. 4. The life cycle of *Tokophrya infusionum* St. The normal adult (upper left) proceeds by budding to form a free-swimming ciliated larva (bottom) which in turn attaches itself and loses its cilia. (From Y. Guilcher.)

budding, the kinetosomes line up in regular rows at one region upon the surface and then this area is delimited by creases that sink in from the exterior. As the bud is being carved out by the ectoplasm, a portion of the large irregularly lobed macronucleus is pinched off to become the macronucleus of the offspring.

Micronuclei divide mitotically and one complement also enters the bud. As the bud detaches, the symmetrically arranged kinetosomes sprout cilia and the free-swimming larva wanders off to a new location. It is this larva which reveals the ciliate ancestry of suctorians. Eventually the larva settles, a stalk is formed as the protoplasmic mass is raised from the substratum, the cilia are lost, and the kinetosomes again assume their haphazard distribution. Finally the small feeding tubes develop at the surface so that the animal can grow and bud in turn.[1]

The case of the suctorians is of particular interest because, though they have a true cytoplasmic development, they are derived from ciliates in which the cytoplasmic structure is to a large extent passed on directly from one generation to the next. Suctorians have retained conjugation and do not use the cytoplasmic development to allow gamete production; but they do use it to produce a small, motile asexual reproductive body. We might presume that there is a selective advantage for small motile buds in this sessile animal and that therefore the extensive cytoplasmic development of suctorians has been secondarily derived. Its similarity to the development of other multicellular organisms would be a case of convergent evolution resulting from the common pressure of natural selection.

Turning now to the description of a quite separate origin of development, let us examine the plant filament. It is easy to make the reasonable assumption that the filament arose from cells which divided but failed to separate. The cells both lost their motility and became encased in a hard polysaccharide wall of a cylindrical shape. The cylinder, besides having obvious advantages of strength, can retain this strength and grow at the same time. From an engineering point of view it is of an efficient design, a matter developed and stressed by a number of the nineteenth-century botanists.

Concerning the nuclear arrangement of filaments, in almost all instances there are small nuclei that divide mitotically. The variations are that the filament may be divided by cross-walls so that each cell is uninucleate, or there may be occasional cross-

[1] For further details, see Y. Guilcher, *Ann. Sci. Nat., Zool.*, (11e série), **13**, 33–132 (1951); and A. Lwoff, *Problems of Morphogenesis in the Ciliates.* Wiley, New York (1950).

walls giving a series of large multinucleate cells, or the whole plant may be without cross-walls to form one large coenocyte. In all three cases the separation of the germ and the soma is simple and straightforward: the gametes are uninucleate and the vegetative portion of the plant increases by mitotic divisions along with cytoplasmic growth and extension of the cell walls.

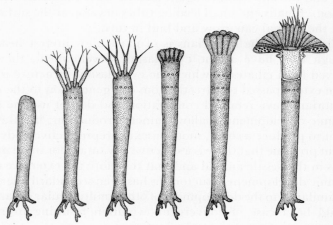

Fig. 5. Diagrammatic representation of the development of *Acetabularia*. Note the formation of successive whorls of sterile hairs and finally the formation of a fertile disk. During the growth stages there is one large nucleus lodged in the rhizoid, but finally at maturity this breaks up into small nuclei each one of which becomes incorporated in a cyst and lies in the fertile disk, as indicated in black in the right-hand drawing. (Based on drawings of Oltmans, Egerod and G. M. Smith of several species of *Acetabularia*.)

An interesting partial exception is the case of the siphonaceous alga *Acetabularia* which has received so much attention of late through the pioneer studies of Hämmerling.[1] In this plant, which is a close relative of the more usual coenocytic algae, there is one large nucleus lodged in the rhizoid, which controls all the cytoplasm, some of which may be lying as far as five centimetres away in the cap (Fig. 5). This nucleus which enlarges as the young plant grows from the embryo is clearly a mass of many genomes; it is comparable in some ways to the macro-nucleus of ciliates. But it has the problem of having to function

[1] For a review see *Internat. Rev. Cyt.* **2**, 475–98 (1953).

as a micronucleus as well and contribute towards the production of gametes. It does this, as the period of sexual reproduction approaches, by breaking up into many small normal nuclei which wander with the protoplasmic streaming into the cap and become, in a series of steps, transformed into uninucleate motile gametes. Meiosis takes place just prior to gamete formation, that is, in the single diploid nuclei. Upon fusion of the free-swimming gametes the diploid zygote settles on the ocean floor and the new plant again develops one large multi-genome nucleus. One cannot help but speculate as to the selective advantage of one large basal nucleus and the answer might lie in the wonderful powers of regeneration that were revealed by Hämmerling. If the plant is sheared off by tides or waves (and the upper part of the plant does regularly disappear during the course of the winter), then it may be advantageous for the plant just to synthesize cytoplasm and not be concerned with making more nuclear material as well.

As the size increases in filamentous forms there is, because of the rigid cell wall, nothing comparable to the morphogenetic movement of animals. This means that, except for protoplasmic transfer in fungi, all the major shaping mechanisms work by growth. The main interest, of course, is how the growth is controlled so that a particular shape appears, but as in each of the examples given here, we shall have to refrain from discussing mechanisms since that is the subject of the next lecture. It should be pointed out, however, that in general there are three main types of construction. There is the coenocytic type which we have already mentioned, in which the protoplasm is continuous (or in large blocks) and the many nuclei are wandering about freely in these large cells by cyclosis, while the cell yet has a definite and intricate pattern. There is the unicellular type found in higher plants in which the uninucleate cells are building blocks and the growth pattern is entirely reflected in the direction and extent, in different regions, of the division and expansion of these blocks. Finally there is the type which might be called 'growth by aggregation' which is found among both the algae and the fungi. In this type the filaments may be unicellular or multicellular. In either case they grow as discrete units; yet their growth pattern is so co-ordinated that the whole mass of filaments has a unified form. The extreme case is that of a mushroom where all the filaments make up a smooth and delicately carved, compound fruiting-body.

As already discussed, one characteristic of many of the large fungi (Ascomycetes and Basidiomycetes) is that fertilization is accomplished by hyphal anastomosis and the flowing together of numerous nuclei. One might wonder whether, in some cases, growth by aggregation arose in conjunction with this kind of fertilization by association and fusion. The intertwining and interaction of hyphae, seen so clearly in the formation of an anastomosis, might have led to more and more efficient systems of association (Fig. 6).

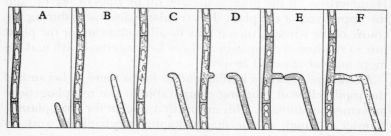

Fig. 6. Anastomosis of hyphae in *Coprinus*. Note that there is evidence here of action at a distance and a mutual co-ordination between the two filaments. (From A. R. H. Buller.)

Turning now to organisms which show some degree of obvious morphogenetic movement in their development, there first is the somewhat rudimentary case of the related green algae, *Hydrodictyon* and *Pediastrum*. To illustrate their cycle let us choose *Pediastrum* because from the recent work of J. G. Moner[1] we have some new insights into its development. A cell from a colony will, under certain conditions, cleave by successive bipartitions giving rise to uninucleate cells of the number 2^n, rarely exceeding 64 (Fig. 7). The mother cell now bursts free from the old colony and each one of the small daughter cells becomes flagellated, swimming freely about in a vesicle. If these are gametes, they escape and fuse to form zygotes, but if they are asexual, then they remain within the vesicle. After a period of swarming the asexual

[1] *Biol. Bull.* **107**, 236–46 (1954); Thesis, Princeton University (1953).

zoospores come together in a flat plate, their movements slowly decreasing until they have assumed correct positions. Then in a surprisingly short time the cells lose their flagella and become radially lobed. Moner was able to show that the disc formation was probably a result of selective end-to-end attachment of the ellipsoidal swarmers coupled with some confinement by the vesicle. Since the organisms are photosynthetic, they now take in energy and grow, ready for the next cycle.

Moner also showed that as the cells grow they produce a substance which stimulates cleavage. If, therefore, the colonies

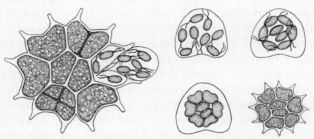

Fig. 7. Colony formation in *Pediastrum*. At the left a mature colony undergoes division in some of its cells which are then liberated in a vesicle (right) and the swarmers proceed to form a flat plate which becomes a new colony. (Based on drawings of G. M. Smith and J. G. Moner.)

grow in a large volume of water, or if the water is constantly changed they will not cleave but grow to a large size. Small quantities of water containing many cells soon reach a threshold concentration of the swarming substance and reproduction begins. The selective advantage of such a system might be that, as a pond dries up, reproduction is stimulated and the product of sexual fusion is a hard-walled resistant stage that could readily survive a drought. The relation between sexual and asexual development and what controls the channelling of reproduction into one of these two directions has not as yet been worked out.

The swarming movement is only morphogenetic or formative in that at the last moment the cells settle by sticking to one another. In the asexual development of *Hydrodictyon* they settle by plastering themselves against the inner wall of the mother cell (Fig. 8). In either case the extent of the control is limited and of a simple nature, dependent upon factors such as the localized stickiness of the cells and the shape of the surrounding wall or vesicle. Its origin is easy to imagine: the inability of flagellated swarming cells to escape the mother cell. Simple as its origin might have been, the resulting shapes are of considerable delicacy and complexity.

In the main animal line, as well as in sponges and *Volvox*, the morphogenetic movements differ from the movements of *Pediastrum* and *Hydrodictyon* in a number of significant ways. First of all the movements are by mass amoeboid motion of many cells, or protoplasmic streaming within the fertilized egg, rather than by ciliary movement. More important perhaps is the fact that in animals and Volvocales the morphogenetic movements are both highly controlled and co-ordinated and play a major, essential part in the morphogenesis of the organisms.

There is no point here in entering into a detailed description of the morphogenetic movements of all these forms. Not only would this be a burdensome and lengthy process, but one which, at least in outline, is familiar to most biologists. I will confine myself to a few general remarks. Morphogenetic movements do not occur at one stage only, but may occur at many different stages. It is true that characteristically the principal movements occur early; oöplasmic segregation takes place just after fertilization, and the all-important movement of gastrulation comes fairly near the beginning of development. Later the movement tends to be more restricted and localized. Furthermore, it should be mentioned that in considering the Volvocales, sponges and all the diverse groups of animals, there is a tremendous variety in different species as to when the movements occur, the type of movement, and the number of different movements.

Concerning their nuclear behaviour, there is great similarity among animals to what is found in the evolution of the filamentous forms. Again the basic presumption is that in the beginning there was growth by mitosis of unicellular organisms that failed to separate and remained colonial. The *Volvox* line from simple aggregates such as *Pandorina* to the complex *Volvox* itself has, for many generations of biologists, remained a tempting idealized picture of the rise of multi-cellularity. Some, such as Dobell,[1]

[1] *Arch. Protist.* **23**, 269–310 (1911).

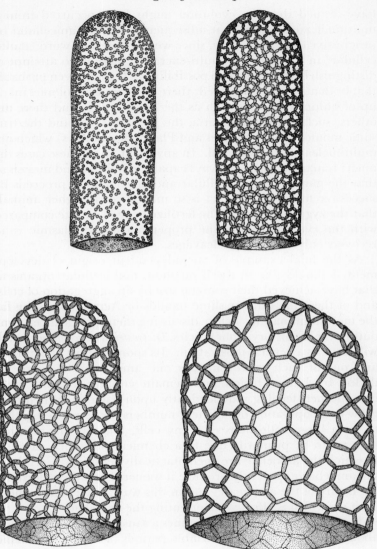

Fig. 8. The growth of a new colony of *Hydrodictyon*. At the left the swarmers have settled down on the inside of the mother cell wall and the growth and elongation of these original swarmer cells produces a new colony which soon bursts free of the old wall.

have argued that the evolution might have occurred from a multinucleate ancestor that subsequently became unicellular by progressive cleavage. Here, since we are using the word 'multi-cellular' in the sense of 'multi-energid', there is no attempt to distinguish between the two possibilities. It seems even probable that both may have happened; there are primitive colonies made up of uninucleate cells such as the *Volvox* series, and there are others, such as the Radiolaria, the Foraminifera, and the true slime moulds (Myxomycetes and Plasmodiophorales) which are multinucleate or plasmodial. In any event, in all these cases the nuclei remained small—always capable of mitosis and meiosis so that the gametes are unicellular and development proceeds by successive mitoses. This fact is so magnified in higher animals that the zygote nucleus of the fertilized egg is minute compared with the cytoplasm, and the proper nucleo-cytoplasmic ratio recovers only after many cleavages.

As the final example of an independent origin of development, I should like to dwell on those multicellular organisms that have achieved their somatic size by an aggregation of cells, and of these the cellular slime moulds or Acrasiales are by far the most illuminating. Before discussing their significance let me outline the life history of one species, *Dictyostelium discoideum*, which will serve as a general illustration. Its spores consist of capsules, and out of each spore emerges one amoeba. The amoebae divide by binary fission and remain entirely separate from one another, feeding independently upon bacteria. When they grow to a population of sufficient numbers they stream together to form large collections of many cells, and this aggregation appears to be primarily due to a chemical substance, acrasin, to which the amoebae are chemotactically sensitive. Certain of the amoebae apparently produce it sooner than others, forming the centre of the aggregate, and in this way an acrasin gradient is set up which is effective in orienting the amoebae.

The aggregated cell mass assumes a sausage shape and crawls about the substratum for variable periods of time, and during this migration phase it is sensitive to light and heat gradients, orienting towards light and towards warmer regions. Differentiation begins at this stage, and the anterior cells are destined to become part of the supporting stalk and the posterior cells will turn into spores. The final fruiting involves a series of morphogenetic movements in which the anterior presumptive stalk cells are pushed down through the spore mass, and in doing so these stalk cells become large and vacuolate and permanently trapped in a delicately tapering cellular cylinder, During this process the spore mass is lifted up into the air and each amoeba in the mass becomes encapsulated in a spore (Fig. 9).

If we now consider these Acrasiales from the evolutionary point of view, it is not unreasonable to suggest that their origin is to be found in the free-living soil amoebae. Many of these free-living amoebae have a resistant cyst stage, which is similar to the encapsulated spores of *Dictyostelium*. Let us assume, and I shall now build a castle of assumptions, that certain biochemical mutants such as deficiencies which kept them from forming cysts, arose in some soil amoebae. Cysts are of obvious selective advantage in their resistance to adverse conditions. However, if two amoebae were together and one were the wild type, then this deficiency might be made up by diffusion of key substances; or if both amoebae had different deficiencies they could, provided they were in close association, make up for one another's faults. Once this condition was achieved they would be totally dependent upon each other for encystment; and note that this hypothesis is really the very same one advanced by Beadle and Coonradt[1] for the first steps of the evolution of sex in the fungi. It is especially interesting to consider the possibility that this advantage in masking deficiencies might actually have been the cause of the origin of some multicellular forms.

The association of cells makes up for the deficiencies, and once the association is established the amoebae are dependent on it for the completion of their cycle.[2] But now a new feature appears, namely, the possibility that there may be some selective advantages to such association other than the mere fact that without it they could not sporulate. It is impossible to ascertain exactly what these advantages might be, although there are a

[1] See page 8.
[2] It should be pointed out here that Sussman (*J. Gen. Microbiol.* **10**, 110–20 (1954)) showed that if certain combinations of ultraviolet-induced slime-mould mutants were mixed, they could often develop further together than either could in an isolated condition.

number of possibilities, and in the next lecture I shall present some speculations.

Assuming selective advantages of the association, we have among the cells a selection pressure for the ability to come together, and we may say then that the whole acrasin-chemotaxis mechanism is the result of such a pressure. Likewise there has been a selection pressure for differentiation, that is, for the production of a stalk which can lift the spore cells into the air. So in each cycle of this mould there will be an internal selection in the cell mass for those cells which are strong in their associative and differentiative potential.

But how does variation occur in such an aggregating organism and how is it transmitted? Variability as usual must be recorded in the genes on the chromosomes of the cells. The production of variation is by mutation and recombination, but so far there has been no convincing or unequivocal demonstration of sexuality, meiosis or fertilization in these forms, although of course eventually some sex mechanism may be discovered. It is interesting to note that the amoebae of *Dictyostelium* before and after aggregation have seven arms to their chromosomes and therefore are no doubt haploid. If there is no sexuality, then a permanent haploid would be most favourable since mutation is to be the sole source of variability.

The success of these mutants depends primarily on their ability to associate and therefore any mutant will be under the immediate, direct selection pressure of success in the fruiting body. This must include the ability to aggregate as well as the ability to differentiate into a spore or stalk cell. Every amoeba must have both these differentiative abilities, for stalk cells die and are not passed on to the next generation; so, if a cell could differentiate into a stalk cell only, it would be eliminated in one cycle.

If, on the other hand, the mutation involves the free amoeba or spore stage, then its survival will not be affected by internal conditions of the cell mass, but solely in its struggle with other spores or amoebae. Such mutants might concern spore survival or feeding characteristics, adaptations which would clearly be affected by changes in the environment. Since the associative phases serve as a gathering of cells of possibly totally different

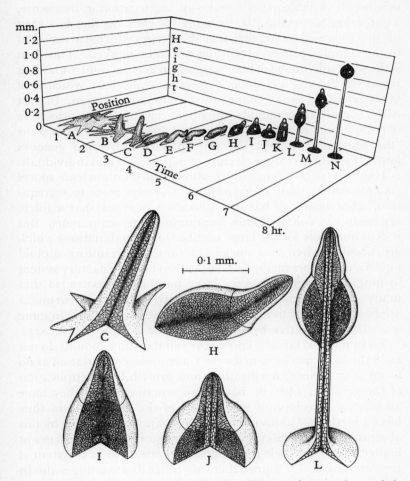

Fig. 9. Development in *Dictyostelium discoideum*. The complete morphogenesis is represented in a three-dimensional graph. A–C, aggregation; D–H, migration; I–N, culmination. The presence of prespore cells is indicated by a heavy stippling, H–K; and the presence of true spores by solid black, L–N. *Below:* semi-diagrammatic drawings showing the cell structure at different stages. The letters indicate the corresponding stages given above.

origin and variable genetic make-up, aggregation is, in a sense, a gathering of variants. By bringing the cells together in a kind of 'heterocaryon' state, deficiencies in genes which are involved in the complex process of fruiting would be masked.[1] This then is the same advantage of heterozygosis in general, for by masking recessives there is a cushion, a reserve, a plasticity which is of great assistance to any change by selection or mutation. In part of the cycle of these moulds (the single-cell stages) the principles of population genetics would seem to apply, while in other parts (the associative stages) the principles of individual genetics apply; they are a curious mixture of populations and individuals.

Haldane[2] says of fungi which also produce uninucleate spores (microconidia) that 'spore formation always leads to segregation. Heterocaryosis has this advantage over sex that a single organism can contain genes from more than two parents. But it does not allow for the large number of recombinations which are possible when two nuclei fuse and the resulting diploid nucleus undergoes meiosis. It appears to be a satisfactory system in fungi which produce very large numbers of spores, so that many types of heterocaryon can be formed.' This argument might apply equally well to the slime moulds and it shows that in some ways the aggregative behaviour could be a substitute for sex.

To sum up, so far as is known the cellular slime moulds do not have the same mechanism for the transmission of variation as do forms with meiosis, fertilization, and growth by multiplication of the genome. They do have gene mutations, and they have an effective mechanism of dispersal of the variants, and they have a method of imitating heterocaryosis. So they have by this aberrant type of life cycle imitated some of the advantages of higher forms, but their great omission appears to be a system of recombination. Of course, there are other disadvantages also in this type of cycle, and there are probably a number of reasons to

[1] M. F. Filosa (Ph.D. Thesis, Princeton Univ. 1958), has some recent evidence which substantiates this view, namely, that various strains in our laboratory are found to be made up of cells of diverse genetic constitution, although the phenotypic effect is characteristic of the wild type only. This *cellular* heterogeniety has been termed 'heterocytosis' to emphasize the parallel to the *nuclear* heterogeniety of heterocaryosis.

[2] *New Biology*, No. 19, 7–26, p. 20 (1955).

explain why so few colonial forms have evolved this particular approach to multicellularity and complexity.

But the pertinence here of these moulds is that in their cycle, from aggregation to the final fruiting body, there is a gene-controlled development, a development that in many ways shows remarkable parallels to that of higher organisms, yet one which does not spring from a fertilized egg, but from an aggregation of possibly genetically heterogeneous cells. When, in the next lecture, we consider the mechanisms of morphogenesis, we will be able to compare, with some constructive benefit, the way in which multicellular organisms of totally different origins have independently achieved the same goal.

THE FUNCTION OF DEVELOPMENT

THE main theme so far has been the inevitability of development, considering the idiosyncrasies of evolution and natural selection, and I have used the variety of different kinds of development that have arisen to emphasize the point. In each case the key was to be found in the twin requirements 'reproduction and size', for the unit of transmitting and cultivating variation has to remain small while selection favours a large somatic individual. It is in effect a simple paradox; for to become large an organism must remain small and the result is a kind of alternation of generations, a separation in time of two processes: one is the step concerned with variability production in the genes and their transmission; the other is the period of efflorescence of the variations, the period of gene action.

In this lecture we shall dwell on this efflorescence, struggling for some deeper insight into its mechanisms, but in doing so we shall not lose sight of its evolution. In fact the thesis maintained here is that through an evolutionary view it is possible to obtain a clearer and more balanced understanding of the workings of development.

The teleological purpose of development is to create an individual that is well adapted and successful in its environment. This rather obvious statement has a number of implications. In the first place there is the fact, already mentioned, that repeatedly during the course of evolution an increase in size has been encouraged and the result, from the point of view of development, is an accumulation of protoplasm by growth. Another implication is that there has been an increase in division of labour, for clearly this is vital to adaptive success. In terms of development this means a differentiation. Increase in size and division of labour then are two primary goals of development, but along with these two there is an equally if not more important goal, although in a sense it is more vague and abstract.

In large measure the success of organisms in nature depends upon their being well knit and closely co-ordinated. It is not enough to have a large organism which parcels out its activities; it must be a discrete, smoothly functioning unit, physiologically well balanced within, as well as adjusted to its environment. It must have systems of physiological regulation, correlation and control so that it is stable, even in adversity. This kind of stability by co-ordination plays an important role in the survival of the fittest and must be constantly improved by selection. But it can only be manufactured in development, and therefore the selection pressure will be towards making the formative processes of development co-ordinated. This is the only way the end result may achieve these attributes. One of the marvellous things about development, which we shall dwell upon shortly, is the high degree of co-ordination and control that goes with it, and now we see that from an evolutionary view we could hardly expect it to be otherwise.

Turning to the solution of the problem of how organisms achieve their goal of a functioning, discrete adult we may begin by considering what factors are present in the germ (be it an egg or a spore) which determine the course of development, for clearly all the directions on how to proceed are present in the beginning. The genes in the chromosomes and the particles and perhaps even the structure of the cytoplasm all contribute in some way. These matters have been discussed in detail elsewhere recently, and here I shall merely outline a few points.[1]

The more difficult subject, for we know least about it, is the contribution of the cytoplasm. There are those cases, already discussed, of clear-cut cytoplasmic inheritance that appear to control qualities of the cytoplasm. Some of these cases may be attributed to specific particles, and others not, but in any event the cytoplasm has many known visible and discrete structures such as (to list a few) mitochondria, plastids, fibrillar networks, kinetosomes, and microsomes of various sorts. These are all directly passed on in each egg and in each cell division. If one considers the matter on a chemical level, there is the great

[1] See C. H. Waddington, *Principles of Embryology*. Macmillan (1956).

multitude of cytoplasmic proteins, fats, carbohydrates, nucleic acids, etc. It is at the moment well beyond the scope of chemical embryology to mark out any kind of chemical epigenesis, but it is plain that the cytoplasm not only contributes in its own right but, even more important, it contributes in conjunction with the genes. The genes act upon the cytoplasm and their action is governed by the condition of the cytoplasm. It is a well-accepted notion that there is not just an action but an interaction between nucleus and cytoplasm.

Gene action has been studied in great detail in recent years, but our knowledge remains rudimentary. One approach has been to study the morphological effect of certain mutant genes or combinations of such genes, attempting to observe the nature of the primary modification during development and how it affects the subsequent embryology. Another approach, used with special effectiveness on *Neurospora* and other micro-organisms is to examine the genetics of synthesis of particular substances, and it is found that certain mutant genes will be instrumental in altering specific enzymic steps in a chain of synthetic reactions. The interest in the field of gene action has recently been so intense that the number of experiments, the number of hypotheses and the number of review articles is depressingly large. Here I will only mention a few notions that will be helpful in our train of thought.

Some geneticists suggest now that we should no longer think of a gene as one discrete package like a glazed bead on a string, but rather that on the chromosome there is a matrix of atoms and molecules (to use an expression of Goldschmidt[1] who has long championed this view) that can act in different ways, depending perhaps on how large a segment is operating at one time. The need for some such concept comes particularly from our knowledge of pseudo-alleles where what was previously thought to be one locus can actually be subdivided by crossing-over.

The genes may not act singly, but often exhibit striking powers of interaction. For example, there are innumerable instances of multiple factors where many interact together to produce a character. In some cases the kind of interaction is quite specific,

[1] *Theoretical Genetics.* University of California Press (1955).

and there may be supressor genes which eliminate the effect of other genes, and related to these are modifier genes which reduce certain specific gene effects. It is also possible to regulate the gene dosage, where certain genes may for various reasons be repeated and this increase in the number of the genes will have various results, either of increasing or decreasing a phenotypic effect. As a final example of gene interaction it should be pointed out that some genes affect the mutation rate of others.

This complexity is not confined to the nuclear apparatus alone, but is seen in every facet of the action of genes upon the cytoplasm.[1] The old simplified notion of one gene being responsible for the formation of one enzyme has not as yet been demonstrated and now the emphasis is rather on the idea that genes affect many cytoplasmic constituents and are likely to be multiple in their effect, so that a discrete event which one would originally have imagined to be the single action of one gene may in fact be the result of a complicated series of interactions involving different parts of the chromosomes and different parts of the cytoplasm. If this turns out to be the case, then the outcome, it must be admitted, is rather discouraging from the point of view of elucidating development in terms of the gene theory and one is tempted to put the matter momentarily to one side and wait for further insights from future experiments.

However, before leaving the subject, there is one relatively simple aspect of gene action, for which we are mostly indebted to R. Goldschmidt, that remains useful and helpful. This is the idea that genes exert their effect primarily through the control of the rates of processes. In a sense it is not surprising that this should be so since genes control enzymes and enzymes are catalysts which control the rates of chemical reactions. Of course this is not the only effect of genes, for by mutation one can have a loss of an enzyme. But by changing the speed of an event, genes can have far-reaching effects on development as Goldschmidt showed. There is, for instance, his case of sex determination in the gypsy moth where the deciding factor is the relative speed at which the set of genes for maleness and the set for femaleness produce their effects. Another example would be the work of

[1] See R. P. Wagner and H. K. Mitchell, *Genetics and Metabolism*. Wiley, N.Y. (1955).

37

Ford and Huxley[1] on the shrimp *Gammarus* where the eye colour (black or red) is determined by the speed at which the genes allow the black pigment to cover the red. Other examples involve growth rates, and therefore the size of many organisms or the size of parts of an organism are subject to such gene-controlled rate processes.

The rate does not merely involve the speed of the reaction, but also the time at which the reaction starts, and for information on this topic we are indebted to J. B. S. Haldane.[2] He has given a number of examples of gene effects appearing at different stages such as in the gametes, in the fertilized zygote, in the endosperm in plants, in the embryo and right up to and including the adult. The effects may even be carried over a generation so that they are not evident until the production of germ cells or the subsequent fertilized eggs, or their effects may appear in the maternal structures associated with the next generation such as the seed and fruit in plants or the egg shell and albumen in animals.

This is another indication then that development is a stepwise process, and it shows specifically that the genes as well as the enzymes are involved in the series of causes and effects. The genes should never be thought of as an initial stimulus which then subsides, but the changes in the cytoplasm interlock with the changes in the genes so that the steps involve both systems and not just the cytoplasm. For instance, certain genes will not have their actions until late in the development, because their immediate environment is such that for a long period they produce no detectable effects. However, when a certain specific set of conditions is achieved through the actions and reactions of other processes, then finally the favourable internal environment appears that allows the genes to bring about some specific event, which, in turn, stimulates or permits succeeding events to occur.

Let us pause for a moment and reiterate the problem that needs solving. In development we have a system which (under rigid

[1] Roux's *Archiv. f. Entw. Mech.* **117**, 67–79 (1929).

[2] *Amer. Nat.* **66**, 5–24 (1932). Also another striking demonstration of this point is the work of Goldschmidt on phenocopies, where the shock treatment produces the phenotypic changes at specific critical periods during the life history.

control) grows and differences arise in different parts. These
differences may be in the amount of growth, in the coarse, over-
all shape, in the detailed pattern, and even in the fine structure
of the cells.[1] The idea of differences in the various regions is
fundamental for the control of growth, of morphogenetic movement, and obviously of differentiation. The central occupation
or activity of development, therefore, is this parcelling-out of
specific materials to specific regions, or, to put the matter
succinctly, *the function of development is the establishing of differences
in different parts*.

There are two general methods by which differences in
different parts are achieved, and most organisms have both,
although one may be emphasized more than the other in a given
species. One is the segregation of units, usually by cleavage.
These units are generally cytoplasmic, although there is a con-
tinued interest in the possibility that mitosis might be unequal
in the nuclei and create a nuclear segregation. Ever since the
discovery, in the embryology of *Ascaris,* of a difference in the
chromosome behaviour of certain cells, this idea has been an
object of speculation and experiment. But even in the case of
Ascaris, as Boveri[2] showed by centrifuge experiments, the behaviour of the nucleus is influenced by the surrounding cytoplasm, again emphasizing the importance of the latter. There
are other cases, however, which leave open the possibility of a
direct nuclear effect, such as the recent work of King and Briggs,[3]
where they show, by nuclear-transplantation methods, stable
differences in the nuclei of frog gastrulae. In any event for both
nucleus and cytoplasm the mechanism is the same: a direct
sorting-out of materials during growth. This is the characteristic
of so-called mosaic embryos. What is left unexplained is how the
direction of this segregation is controlled so rigidly and consis-
tently from generation to generation.

The other method of achieving differences is by having the
units represented in all the parts, but here they are called forth in
specific ways in specific regions so that the regions themselves

[1] See Waddington, *Principles of Embryology* (particularly chapter xx) for
an excellent discussion of the elements of development.
[2] See E. B. Wilson, *The Cell.* 3rd ed. Macmillan (1928).
[3] *Proc. Nat. Acad. Sci.* **41**, 321–5 (1955).

become different. This is characteristic of so-called regulative development. Here the unexplained aspect is the control of the 'calling forth': what determines the regional distribution of the mechanisms which bring out the differences in the homogeneous mass of protoplasm (homogeneous, at least, as far as its potentialities are concerned)?

Again it must be emphasized that these two methods undoubtedly exist to some extent in most if not all organisms. That is, there is usually a period when all regions are equipotential and this is followed by a fixation of parts or a 'determination' as it is called, and then each region has taken on a specific character. As was noted, both periods need further explanation of their method of control of spacing: in the first the spacing systematically calls forth specific characters in specific regions in a protoplasmic mass containing all the determinants in all the parts; this is followed by a segregation of the determinants into different regions, which is again a major problem of spacing.

The nature of these spacing mechanisms is a most interesting aspect of embryology. It is one about which there remains great ignorance, but we do have some information. The prime purpose of these controls is to preserve the unitary character of the developing embryo. In order to do this they must command the whole structure; they must transcend the cells or the energids and spread through whole regions; they must allow some sort of communication between parts so that one part knows what the other is doing and can act accordingly. It must be a harmonious co-ordinating system.

We can roughly classify these systems into a number of categories: gradient systems, induction and polarity. There may of course be other systems which are as yet unknown, and even our knowledge of these three is fragmentary. I should now like to discuss each one in a few words, preparing the way for an analysis of their operation and evolution in a number of divergent types of organism.

Gradients are known to exist in some form in all types of embryo that have been examined. Child[1] devoted a life study to them, although his emphasis was primarily on the metabolic aspects of gradients. We tend now to think in the more general

[1] *Patterns and Problems of Development.* University of Chicago Press (1941).

terms of a gradient of chemical components, which in many cases may be reflected in a respiratory or metabolic gradient. Nevertheless, Child contributed greatly to our understanding of how gradients could arise, and how they can be effective as correlating mechanisms.

The most obvious method of the origin of a gradient is the direct effect of the external environment. If a mass of tissue is resting on the bottom the underside obtains less oxygen and accumulates more carbon dioxide than the upper surface, and the result will be a chemical gradient. There are many other possibilities of a similar sort, for example, the position of the egg in the ovary may give rise to a gradient as a consequence of its location. If the environment does not have this differential effect, then there are possible internal causes of gradients. There are numerous hypothetical schemes to illustrate this point, the simplest of which is perhaps the idea of competition for a particular substrate, and once a successful reaction (which we shall assume is at one end of the cell mass) begins it will grow and drain the substrate from other competing reactions, which will result in a gradient.[1] The mathematical ideas of Rashevsky[2] and Turing[3] contribute useful hypotheses for such mechanisms, as have also the many interesting studies on adaptive enzymes.

Child argued with considerable force for the quantitative differences along a gradient resulting in qualitative differences. If one makes, for instance, the reasonable assumption that thresholds play a part, then once a certain threshold is achieved in a particular region along the gradient, another local reaction may begin which will change that region in a specific way. That local environment may be such that a particular set of genes may begin to act which then may lead to further, more detailed, local changes.

The kind of gradients just outlined would explain control in regulative forms where the units are distributed throughout the developing organism and different parts are called forth to produce qualitatively different structures. Gradients could also play a role in the control of mosaic development. Imagine an egg

---

[1] This is the basis of Child's concept of dominance.

[2] *Mathematical Biophysics*, rev. ed. University of Chicago Press (1938).

[3] *Phil. Trans. Roy. Soc.*, Ser. B, **237**, 37–72 (1952).

which is placed in an asymmetrical environment; the result is that an internal gradient is set up. Then with the advent of cleavage these differences are segregated in the different blastomeres.

The process of induction is rather a special case, but because of its widespread importance it deserves attention. In order to have induction one must have a previous distribution or regional differentiation. The stage must be set so that there is a stimulating region presented in close proximity to a region capable of responding. This preliminary spacing, which may arise either by the regulative or the mosaic method (or more likely a combination of the two) then inevitably leads to the stimulus and the response which in turn leads to further elaborate changes and spacings. Induction emphasizes the epigenetic character of development, for the induction itself is one step in a whole series of steps.

The term 'polarity' is a useful catch-phrase for any directional quality evident in development. For instance, a gradient such as we have just described would be a manifestation of polarity. The importance of polarity is that by some manner there is an orientation within the embryo, and this orientation will be instrumental in the fundamental spacing we have been emphasizing.

In the case of gradients the important associated physical mechanism is presumably diffusion. That is, communication between one part and another is by diffusion of substances. This is by no means the only method. In plants there is a directional or polar movement of the growth hormone, auxin. It will go in one direction far more readily than in the opposite, in certain regions of the plant, and its rate of movement greatly exceeds what could be expected from diffusion. Furthermore, by this polar movement one can accumulate auxin against a concentration gradient. It is not known what physico-chemical mechanism is responsible for polar transport, but its pertinence to the argument here is that substances (in this case an inductor substance) can be moved in a directional fashion by means other than diffusion.

Another kind of polarity observed in embryos is a polarity in the morphogenetic movement. Part of the fundamental spacing (in animal embryos in particular) is the result of the movement of material from one part to another by the amoeboid motion of its constituent cells. But the direction of this motion in itself must be guided and there are two principal methods known. One is the direct result of gradients. The amoeboid cells are oriented in a gradient of a particular chemical substance so that they either go to the point of high or low concentration. The slime moulds, in their orientation towards high concentrations of acrasin, give a good illustration of the former, while the migration of melanophores in the newt *Triturus*, as Twitty and Niu[1] have shown, illustrate the latter.

Movement may also be guided by direct mechanical means. This has been studied in detail by P. Weiss[2] who has shown that by 'contact guidance' as he calls it the amoeboid cells feel their way along grooves and fine structure in the substratum so that they are orientated by the orientation of the substratum itself. Also they are sensitive to small tensions and in this way they will follow one another if they are in close association. It is like the circus elephants, each with the tail of the next in its trunk, and in this way a whole body of cells will move as a unified mass. Holtfreter[3] showed the importance of this tension, follow-the-leader mechanism, in amphibian gastrulation. There all the cells are attached to the surface coat, which by its contraction appears to guide the movements of all the cells.

As was implied, polarity may also reside in the molecular orientation of the fine structure of an organism. In higher plants this is seen in the orientation of the cell walls; in higher animals it is especially evident in such structures as nerve, muscle and tendon; in ciliates it is seen beautifully in all the surface structure of the cell. The mechanism of this kind of spacing is still obscure, for in some ways it resembles straightforward chemical crystallization, and in others it appears totally different and mystifying. This is likely to continue to be a fruitful area of investigation.

To sum up, the basic function of development is the orderly spacing of constituent parts. This spacing is achieved by the communication between parts (e.g. gradients, polarities,

[1] For a review, see V. C. Twitty, *Growth Symposium*, **9**, 133–61 (1949).

[2] For complete references see his most recent paper, *J. Exp. Zool.* **100**, 353–86 (1945).

[3] *J. Exp. Zool.* **93**, 251–323 (1943); **94**, 261–318 (1943); **95**, 171–212 (1944).

inductions) and characteristically occurs in a stepwise epigenesis. In the fertilized egg or asexual spore all the needed components are close together in one small unit. With the expansion that results from growth there are a series of spacing mechanisms, one following another (for the end condition of one leads to the beginning condition of the next), and in so doing a complex, elaborate, and unified individual is the result. It should be

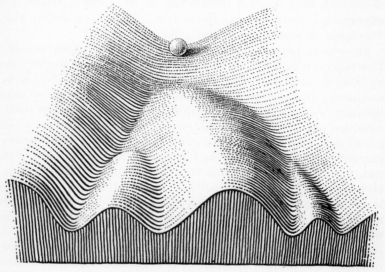

Fig. 10. An 'epigenetic landscape'. A representation of a developmental system as a surface (sloping towards the observer) on which there are valleys along which the processes of differentiation tend to run. (From C. H. Waddington.)

added that one of the characteristics of the steps is a certain degree of specificity so that usually the reactions can take but one course. Occasionally the reactions are set up in such a way that an alternate course is possible (with the appropriate push), but it is either one or the other—there is never any intermediate. This quality has been examined in detail by Waddington who labels it 'canalization' and has illustrated the point by an intriguing drawing of a ball rolling down one of the grooves in an 'epigenetic landscape' (Fig. 10).

If we now turn to specific cases in organisms of different developmental origin, then we can observe the variety of ways the common function of development has been carried out. Again we shall use ciliates as our first example. The fact that in evolution there arose such an unusual and unorthodox development as that of ciliates is particularly useful to us in this inquiry when we compare it to the development of other forms.

The macronucleus of ciliates, which is responsible for controlling the cytoplasmic structure, has the whole gene complement repeated many times in a kind of polyploidy. It has been known for some years that numerous ciliates, for example *Stentor*, will regenerate from small bits provided the bit contains a portion of the macronucleus. Therefore, presumably all the genes are distributed in all the different parts of the macronucleus and the spacing of development in ciliates does not come from spacing within the nucleus. In a sense the macronucleus is a harmonious equipotential system. It is interesting to note that the shape of the macronucleus varies greatly in different species. In *Stentor* there are a series of nodes like beads on a string; in *Euplotes* there is a large C, in *Paramecium* there is an oval-shaped structure, and there are many others. In no case does the shape of the macronucleus reflect the shape of the cell; sometimes the nucleus may be a considerable distance from the cytoplasm that it affects. Also the shape of the macronucleus may vary during the division cycle. For instance, in *Stentor* the beads come together to form an oval-shaped structure at the moment of bipartition, and this oval cleaves in two by amitosis.

Recently it has been shown in some interesting experiments of P. B. Weisz[1] that there is an important exception to the notion of regional equipotentiality in the macronucleus. The old observations hold for long periods after cell division, but eventually as the interphase approaches the next division, certain regions lose their ability to promote complete regeneration in cutting experiments. In *Stentor* the posterior nodes can no longer promote the formation of new mouth structures, but when the macronucleus bunches together during binary fission, the gene material is again distributed to all parts of the new nuclear chains of both daughter cells. The immediate cause of this loss of potency in later interphase was investigated by Weisz and by moving anterior nodes into the posterior region he showed that the effect

[1] Reviewed in *Amer. Nat.* **85**, 293–311 (1951).

on the nucleus came from the surrounding cytoplasm. This leads us quite directly to the fact that the spatial differences in ciliates are found in the cytoplasm rather than in the nucleus.

Spacing in ciliates is a subject of great interest because of the wealth of fine detail manifested by the cilia, the basal bodies, the silver-staining networks, the trichocysts, etc. We owe our basic understanding of these structures largely to the work of Chatton, Lwoff and Fauré-Fremiet,[1] who showed that the most fundamental unit was the basal body or kinetosome; for not only does this body give rise to the cilium, but also it is the unit in the development of trichocysts, contractile vacuoles, mouth structures, membranelles and cirri. Furthermore, Chatton and Lwoff pointed out the fact that these kinetosomes lie in rows (kineties) and that to one side of these rows there is a detectable line (kinetodesmy). The interesting thing is that the line is always on the same side of the row of kinetosomes (their 'rule of desmodexy') so that essentially each little bit of surface has a directional quality, a polarity. If the anterior end of the cell is considered to be the north pole then the kinetosomes will always be east of the line. This means that in a polar view it is possible just by looking at the positions of the kinetosomes and the lines to determine if it is the north or the south pole (Fig. 11). The future progress in the investigation of these surface structures lies in electron microscopy and already Metz, Pitelka and Westfall[2] have shown the structural basis of the rule of desmodexy in *Paramecium*. The line is composed of tapering fibres of equal length that lie side by side and are staggered so that each extends a third beyond the next, like shingles on a roof from a profile view. At the thick end of each of these tapering fibres there is a kinetosome; in fact the kinetosome has a tapering fibre at each end, one extending in the direction of the north pole and the other extending outward in the form of a cilium (Fig. 12). Therefore the polarity of the organism is reflected in each minute segment of the surface.

As one would expect from such a delicately polarized surface, there is no evidence of polarity reversal in grafting or cutting

---

[1] For a review, see E. Fauré-Fremiet, *Folia Biothéor.* Series B, **3**, 25–58 (1948); and A. Lwoff, *Problems of Morphogenesis in Ciliates.* Wiley, N.Y. (1950).

[2] *Biol. Bull.* **104**, 408–25 (1953).

experiments. As Tartar[1] and Weisz[2] have shown, particularly in *Stentor*, a graft with individuals pointed in the opposite direction will either remain stable for some time as a kind of mirror-image twin or the individuals will turn towards each other and fuse in the same direction. In any event no part of the surface reverses

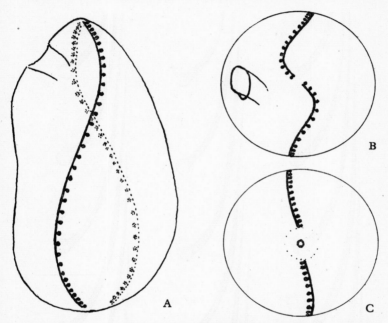

Fig. 11. A diagram of the orientation of surface structures of a ciliate according to the observations of Chatton and Lwoff. A, side view showing the mouth, and one kinetodesmy with the kinetosomes to its right; B, surface view of the anterior end; and C, surface view of the posterior end to demonstrate the rule of desmodexy. (From E. Fauré-Fremiet.)

its direction. The point is also well illustrated in an experiment of Weisz where he cut a large portion (including the mouth region) from a *Stentor* such that when the extended wound closed the rows of kinetosomes went up one side of the animal and down the other (Fig. 13). The new mouth did not form at the top but upside down on the side where the kinetosome rows point down-

[1] *Growth Symposium*, **5**, 21–40 (1941); *J. Exp. Zool.* **127**, 511–76 (1954).
[2] *Biol. Bull.* **100**, 116–26 (1951).

ward, and then eventually, by growth, the individual straightened out.

This superficial, polarized frame is a kind of space lattice and in each cell division it is passed on directly to both the daughter cells. Almost invariably the division line cuts across one elongate

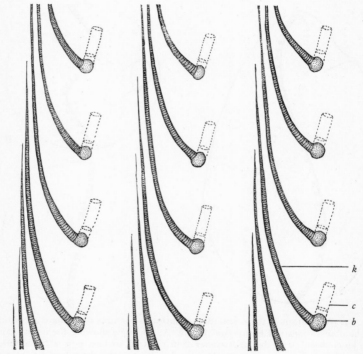

Fig. 12. A diagrammatic drawing of the surface structure of *Paramecium* as revealed by the electron microscope: *k*, kinetodesmy; *b*, basal body or kinetosome; *c*, cilium. (Drawing based on the photographs of C. B. Metz, D. R. Pitelka and J. A. Westfall.)

individual and the polarity is predetermined for both daughter cells.[1] This is really a case of the direct inheritance of a spacing mechanism. Chatton and Lwoff have emphasized the genetic continuity of kinetosomes, pointing out that, not only do they

[1] There is an interesting exception in *Halteria* described by E. Fauré-Fremiet, *Arch. Anat. Micr. et Morph. Exper.* **42**, 209–25 (1953).

give rise to many of the surface structures, but since they can duplicate themselves they constitute a system of cytoplasmic inheritance. To this point we are simply adding the idea that not just the kinetosomes are passed on but their polarity as well.

It was emphasized in the first lecture that the cytoplasm of ciliates had reached such a state of complexity that in a sense it had lost some of its power of development, with the result that

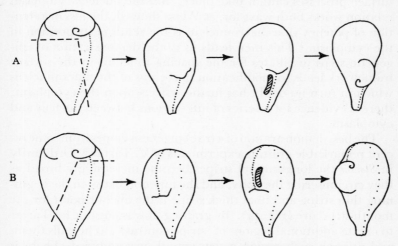

Fig. 13. Regeneration experiments on *Stentor*. The dotted lines are the extent and path of the left boundary stripe. The part of the animal bracketed by the broken lines was removed, and note that in each case the regenerating mouth structures formed at the upper end of the transected left boundary stripe and the orientation of the oral zone was constant with respect to the orientation of the stripe and not with the overall shape of the whole *Stentor*. (From P. B. Weisz.)

conjugation has appeared and there is never a complete reduction of the cytoplasmic structure in the formation of small gametes. We have now seen evidence for the fine detail of the cytoplasmic surface and it is quite understandable how such a space lattice might be, in terms of efficiency, easier to pass on directly rather than to be re-created each generation.

This does not mean, however, that all the spacing for the form of the daughter cells is laid out in the beginning, for in hypotrichs such as *Euplotes* there are well-controlled expansions of surface structures that occur during and after division. How exactly these movements are controlled is a puzzling question; all that is known is that the rate of expansion between parts is equal, which is not inconsistent with the notion that there may be some crystal-like properties in the surface structures.[1]

Again it is not known in what way the macronucleus affects these surface movements or kinetosome behaviour in general, although there is clear evidence that without the nucleus the surface processes cannot take place. But the nucleus-cytoplasm relation works both ways for, as Weisz showed, the loss or retention of potency in a macronucleus is the result of its position in the cytoplasm.[2] This then leads us to the demonstration of epigenetic steps in ciliates for the spacing inherent in the surface framework leads to modification in regions of the macronucleus which in turn no doubt has further effects upon the cytoplasm; there is evidence for a series of interactions between nucleus and cytoplasm.

The best demonstration of a true epigenesis comes from some recent remarkable grafting experiments of V. Tartar.[3] Ordinarily in *Stentor* the longitudinal stripes become increasingly broad as they circumscribe the body, and the junction between the beginning thin stripe and final thick stripe is the anchor point for the mouth structure (Fig. 14). By grafting, it was possible for Tartar to create additional regions of 'stripe contrast', as he calls them, and at each such region a new mouth was induced. There is some quality produced by the juxtaposition of thin and thick stripes which promotes the formation of oral structures.

If we now turn to a totally different group of organisms, such as the higher plants, we shall see that their spacing mechanisms and epigenetic steps are of a somewhat different nature. Plants do have, in common with ciliates, a rigid, oriented surface structure in the form of the cell wall, but this wall which consists of oriented cellulose, is not so much part of the cytoplasm, but rather a deposit, an accretion, an exoskeleton. Cell walls do

[1] J. T. Bonner, *J. Morph.* **95**, 95–108 (1954).

[2] See page 45. The importance of the cytoplasm on the nucleus has also been demonstrated by Sonneborn (*9th Internat. Genet. Congr. Carylogia Suppl.* **6**, 307–25, 1954) who, with his co-workers, has shown that the persistence of micronuclei following meiosis depends upon their location in the cytoplasm.

[3] *J. Exp. Zool.* **131**, 75–122; **131**, 269–98 (1956).

show polarity; it is easily revealed with the polarizing microscope as birefringence and it is the result of the orientation of the polysaccharide fibrilles that make up the wall. It is not at all clear, however, to what extent this orientation serves as a spacing mechanism in development. It is quite possible that the orientation is solely a mechanism for giving strength and that we must look elsewhere for systems of spacing development.

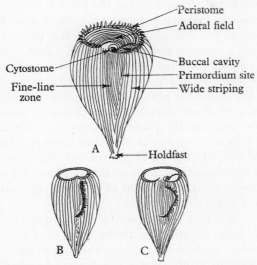

Fig. 14. Drawing showing the orientation of surface structures of *Stentor* under various conditions. A, vegetative stage showing the zone of stripe contrast which is the primordium site; B, the mouth region has been excised and the regeneration primordium arises in the zone of stripe contrast and then later moves up and assumes its anterior position; C, an early stage of normal division where the primordium of the future posterior daughter cell arises in the zone of stripe contrast. (From V. Tartar.)

For the most part, in the cases that have been properly tested, the units of differentiation and development are in all the parts of the developing plant (regulative development).[1] Furthermore,

---

[1] This statement is based primarily on tissue culture studies where small parts of a plant give rise to a large variety of cell types and whole plants will regenerate. Also isolation of parts of the stem apex results in complete regeneration. There is, however, much to be learned in this area of plant development. For instance, so far as I am aware, no studies have been made of plant embryos to determine whether they are mosaic or regulative.

4-2

in higher plants morphogenetic movements are at a minimum and all the pattern is produced by differential growth. If the spacing is not to be found in either the mosaic partitioning of materials in cell division, or in the orientation of the cell walls, it is fair to ask where these spacing mechanisms are, for certainly if we look at the variety of shapes in higher plants there is undoubtedly an accurate and detailed control. The answer is unfortunately hardly complete, but the polar transport of growth hormones plays an extremely important part. It is far from being the only factor, for different regions of the plant respond in different ways to similar concentrations of auxin, and different regions of a plant produce different quantities of auxin. All these differences are, in a rough sense, differences which result from polarities within the plant, but our knowledge of the physics and chemistry of these orientation mechanisms is almost non-existent. We can describe plant development to the extent of saying that it is the regulative type and indicate that polarity plays a vital role in the 'calling forth' of different properties in different parts of the plant, but this is about all we can do, and therefore it is an admission of abysmal ignorance of the mechanism of plant development.

Because of the lack of knowledge it is equally difficult to speculate on how the spacing mechanisms may have arisen in evolution. One approach to this problem will be the analysis of development in multicellular algae and already there have been some fruitful investigations, but it is not even clear in many of these instances how the algal spacing mechanisms correspond, if at all, to those of higher plants.[1]

The fungi, especially the larger aggregative forms such as mushrooms, have one feature which is quite different from the multicellular uninucleate plants: they have a kind of morphogenetic movement in which the protoplasm may flow through the length of the filaments since the cross walls, if they exist, are often perforated. This fact, and the fact that mushrooms are the compounding of many filaments, produce a unique type of development which I shall briefly describe.

[1] Recently P. Green of Princeton University has done some most interesting studies on the green alga *Nitella* where there is little evidence of auxin-like control of dominance and stem elongation.

Upon spore germination the hyphae spread through the soil or compost, sopping up food and increasing in size. The basidiospores are haploid and upon fusion (often of numerous hyphae) a dicaryon results (Fig. 15). Eventually the mycelial mat will

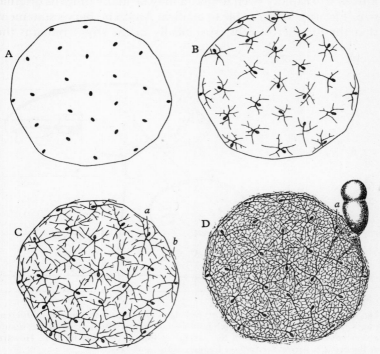

Fig. 15. Communal development in the mushroom *Coprinus*. A, B, spores germinate on a restricted area of nutrient media; C, D, ultimately they anastomose and jointly produce one fruiting body. (From A. R. H. Buller.)

have reached such a size that, provided the atmospheric conditions are favourable, small buds or concentrations of tangled hyphae will appear at the surface of the substratum. These are primordia and some of these will develop into mushrooms. Using histological techniques, it is possible to observe that when these buds are small there is no orientation to the hyphae, but as they enlarge to the height of about 5 mm. (in the common mushroom *Agaricus campestris*) the hyphae slowly assume an oriented position so that in the small stipe the filaments run up and down, and there is a crude radial orientation in the minute cap.[1] The mushroom is now essentially blocked out in miniature and with the proper moisture and temperature, this button will rapidly swell so that it may be many times its original size (Fig. 16). We are able to confirm Anton de Bary's statement that the number of cells (specifically in the stipe) remains the

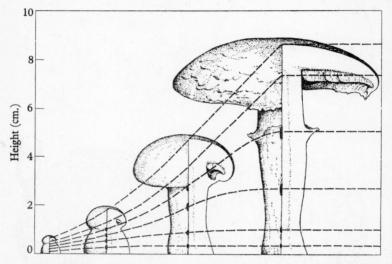

Fig. 16. A diagrammatic growth curve of *Agaricus campestris* showing four different stages. The lines are drawn through homologous points. The horizontal axis is only a very rough approximation of the passage of time. (Drawing by J. L. Howard in the *Scientific American* based on Bonner, Kane and Levey.)

same, and that the over-all expansion can be entirely accounted for by the cell expansion. The expansion is not generalized over the length of the stipe, but as Fig. 16 shows it is localized in the region just below the cap.

By taking dry weights and wet weights of mushrooms at different stages of development it was possible to show that their ratio remained the same, indicating that the increase is not due to a sudden intake of water, but that there must be a surge of material from the deep soil mycelium into the ex-

[1] J. T. Bonner, K. K. Kane and R. Levey, *Mycologia*, **48**, 13–19 (1955).

panding fruiting body. Evidence for such a phenomenon in many different types of fungi was firmly established by Buller[1] (e.g. see Fig. 15), who presented evidence in favour of the movement being caused by osmotic differences between one end of the filaments and the other.

I have dwelt in such detail upon mushroom development because it is one that is unfamiliar to many biologists. As the primordium appears on the surface of the soil there is no reason to suspect any mosaic sorting out of materials, but all the materials are likely, at least in the beginning, to exist in all the regions of the bud. The important, and completely mysterious process, is the orientation and spacing that takes place before the button reaches the 5 mm. stage. Everything appears to be done (in the way of spacing) in this small hyphal mass, when the large bulk of the protoplasm lies deep in the soil mycelium. However it originates, the result is a miniature, predetermined embryo, which like an unfilled balloon, merely needs the proper filling substance to blow it up. This filling is really a kind of morphogenetic movement, for there is a migration of protoplasm; but this case is unusual in that, instead of the morphogenetic movement occurring early and in part being responsible for the shaping of the pattern, it occurs late, after the pattern is laid down, and merely contributes to expansion.[2] But again, as in the case of higher plants, it must be admitted that as far as spacing mechanisms are concerned, we remain pretty much in the dark.

If we look to the origin of mushroom development, we can trace what appears to be a fairly continuous course of increasing compounding of mycelia, starting with simple fruiting bodies such as *Rhizopus* or *Aspergillus*, passing through the coremia forms which are an aggregate of single-filament sporangia (Fig. 17), and then finally passing to the fleshy ascomycetes and basidiomycetes. In each step there must have been a closer communication between filaments, perhaps of the inductive type known to exist in hyphal anastomosis (and illustrated in Fig. 6). In this case one filament will give off what is presumably a diffusible substance which produces a reaction (a positive chemotropism)

---

[1] *Researches on Fungi*. Longmans, Green and Co. (1909 et seq.).

[2] Not all fungi have this type of development, but some continue to show apical growth until the very end of fruiting-body formation.

in a neighbouring filament. There are many other examples of such interactions between filaments, for example the work of D. R. Stadler[1] on the negative chemotropic effect of substances given off by the hyphae of *Rhizopus*. Now it is for the experimental botanist to find out how such free diffusing substances could be so

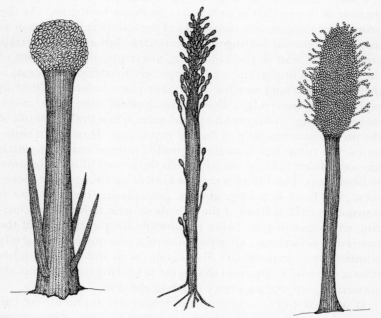

Fig. 17. Three different types of coremia found among Fungi. Left, *Graphium*; middle, *Stysanus*; right, *Trichurus*. (Based on drawings of H. L. Barnett.)

organized and so spaced as to contribute towards the moulding of the small mushroom primordium of, for instance, *Agaricus campestris*.

Higher animals, with all their diversity, have received far more attention than any other group of organisms. For the most part it is a story familiar to zoologists, and if not it is discussed in many places in great detail, all of which leads me to make only the briefest mention of it here—just enough to remind you how

[1] *J. Cell. Comp. Physiol.* **39**, 449–74 (1952); *Biol. Bull.* **104**, 100–8 (1953).

the familiar animal embryology fits in with the other organisms we are discussing.

One of the striking features of animal embryos, seen especially among invertebrates, is the presence of both mosaic and regulative forms among groups of moderately close phylogenetic relation. If we state the matter in our terms of spacing, then there are those cases where the distribution of parts occurs early (mosaic forms) and those cases where the distribution occurs late (regulative forms). The distribution process is a step in a chain of events and the timing of this step is important, for if it occurs early the regulative powers of parts of the embryo will be correspondingly decreased. In both cases the distribution must be controlled, as has already been pointed out: in mosaic forms the control must produce a method of segregating the material that is consistent from generation to generation; in regulative forms the control appears in the 'calling forth' mechanisms that have a regional distribution of their own, presumably as a result of gradients, polarities, etc.

In mosaic eggs there is, usually following fertilization, a redistribution, a flowing of materials within the egg. The result of this early morphogenetic movement, or oöplasmic segregation as it is more properly called, is that there are different substances in different regions that are now partitioned off or parcelled by the subsequent, successive cleavages. During these cleavages there is essentially a cytoplasmic inheritance taking place, for the differences which ultimately arise in the different cells are passed on through the cytoplasm. But the key problem is the control of the oöplasmic segregation, although unfortunately virtually nothing is known of it. The best we can do is to put forward hypotheses, and Costello[1] has suggested as a guiding factor diffusion gradients arising from differential permeability which in turn is the result of fertilization.

Let us examine oöplasmic segregation from the point of view of its evolution. In development a mechanism is needed which distributes cell constituents so that there can be different local environments, assuming that a differentiation, a division of labour, has selective advantages. If some primitive hypothetical organism, consisting of two cells that arose from the division of

[1] *Ann. N.Y. Acad. Sci.* **49**, 663–83 (1948).

57

one, is to follow such a plan, then an inherited method must be devised so that this one cleavage will separate the cytoplasm asymmetrically to produce such a differentiation. This asymmetry cannot be retained at all times for it would interfere with the sexual mechanisms. Therefore it must be brought about rapidly in the short interval between fertilization and the one cleavage. Any external or internal stimulus that will cause the movement of cytoplasmic particles will be retained by selection so that the stimulus and response ultimately become fixed and controlled, and this may well involve, as Costello has suggested, a conditioned dependency upon diffusion forces.

The origin of regulative development, using a similar hypothetical two-celled animal, would be slightly different. Here both cells would have all the cytoplasmic and nuclear characters; the division would be an equal one in every sense. Then, since again we assume the selective advantage of differentiation, one cell, because of its position might become different from the other. If one cell were, for instance, resting on a substratum, there would be (to use the argument of Child) more carbon dioxide and less oxygen in its vicinity and this would trigger a particular set of reactions which would result in some differentiation. Again this response mechanism must be fixed and inherited by selection. Since the cells are equal at division it should be possible to invert the two cells and now the differentiation paths of the two cells would be reversed.

These two crude examples have serious limitations. They fail to emphasize the incredible complexity of the development of the animals that exist today. This complexity includes the fact that both of the mechanisms described occur together in one organism, and that there are many epigenetic steps one following the other. The sequence and the nature of these steps will vary with the species of the animal. The steps involve an orchestrated concert of morphogenetic movements, segregation of substances, growth, gradients, inductions, and many others—all an orderly unfolding of the nuclear genes and cytoplasmic constituents of the fertilized egg.

In discussing gene action, and now here in discussing animal development in general, we arrive at the discouraging conclusion that the more we probe the more we discover complex interrelations. The antidote prescribed by most animal embryologists and experimental botanists is to confine oneself to a single organism, or perhaps even a single step in one organism and attempt to see the problem through. There are many gaps in our knowledge, and much need for hypotheses, but I shall try, in the remainder of this lecture, to analyse the development of cellular slime moulds in this way. As with ciliates, slime moulds have a separate origin of development; yet they have the common function of all development—the controlled spacing of the constituent parts.

Let me remind you that a cellular slime mould arises by the aggregation of many cells and the evidence is excellent that these cells are equipotential. In this case, then, there can be no question of a mosaic segregation—it is an ideal example of regulative development. The dictum of Driesch is strictly adhered to: the fate of a cell is a function of its position within the whole. As the cell mass moves across the substratum, the anterior cells begin their differentiation into presumptive stalk cells and the posterior cells turn into presumptive spore cells. Eventually, in *Dictyostelium discoideum*, the 'sausage' rights itself and a small fruiting body results, in which the anterior cells all turn into the delicately tapering stalk, and each of the posterior cells becomes encapsulated as a spore and is lifted into the air to form an apical sorus (Figs. 9, 18).

There are some related species which make interesting comparisons. In *Dictyostelium mucoroides* (and *D. purpureum*) there is a similar development, apart from the fact that during migration the stalk is continuously formed, from the end of aggregation on to maturity (Fig. 18). In the genus *Polysphondylium* the stalk is also continuously constructed, but it forms many sori instead of one. The cell mass fragments at its posterior end, leaving behind a series of collars of cells which cease forward movement and each of which break up into several radially oriented secondary fruiting bodies giving a series of whorls (Fig. 18).

Using various techniques it is possible to reveal the extent of the presumptive spore and stalk areas during migration but, as shown in Fig. 19, such a difference is only demonstrable in the *Dictyostelium* species; in *Polysphondylium* the cells appear immature or embryonic up to the very last moment before final differen-

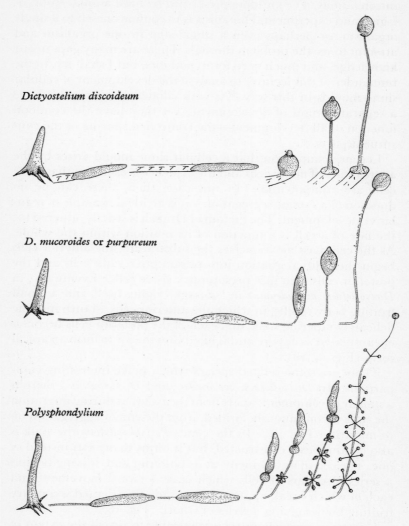

Fig. 18. A semi-diagrammatic drawing illustrating the migration and culmination stages in three different types of cellular slime moulds.

tiation.[1] We have suggested that this delayed differentiation (or slime mould neoteny) in *Polysphondylium* is the factor which permits the breaking-up of the cell mass at the end of migration. In *Dictyostelium* these sharply delineated areas are of great interest because they are proportionate in fruiting bodies of different sizes. Since the size of an individual is dependent upon the number of amoebae that enter one aggregate, there is in any

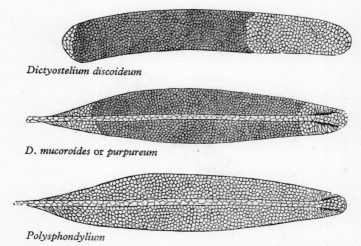

Dictyostelium discoideum

D. mucoroides or *purpureum*

Polysphondylium

Fig. 19. A semi-diagrammatic drawing showing the staining properties of the cells during the migration stage for three different types of cellular slime moulds.

normal population a great range in individual size: there may be as great as a thousand-fold difference in volume between the smallest and the largest cell mass.

This matter of proportion is of such importance because it is the essence of a controlled spacing mechanism. We begin with an equipotential mass of cells, and after a period they take one of two paths of differentiation. There is a sharp dividing line between these two groups and the division always takes account of the total number of cells. The exact nature of the propor-

[1] The details of this work on differentiation in slime moulds will be found in J. T. Bonner, A. D. Chiquoine and M. Q. Kolderie, *J. Exp. Zool.* **130**, 133–58 (1955); and J. T. Bonner, *Quart. Rev. Biol.* **32**, 232–46 (1957).

tionate relation is an allometric one (Fig. 20).[1] When plotted logarithmically, straight lines are obtained; in *D. discoideum* the slope is one, while in *D. mucoroides* it is greater than one. This means that in *D. mucoroides* the anterior wedge of presumptive stalk cells is relatively smaller in large cell masses.

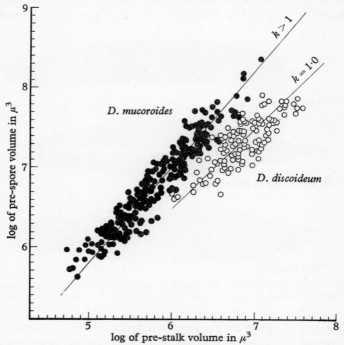

Fig. 20. A graph showing the log of the pre-spore volume plotted against the log of the pre-stalk volume for *Dictyostelium mucoroides* and *D. discoideum*.

It has been known, ever since some pioneer experiments of K. B. Raper,[2] that if a piece of the slug is removed it will regulate and form a small, proportioned fruiting body (Fig. 21). We have followed the process in the cell mass, using histochemical tech-

[1] $Y = bx^k$ where Y and X are the volumes of pre-sorus and pre-stalk respectively, b a constant and k the slope of the line. See J. S. Huxley, *Problems of Relative Growth*. Methuen, London (1932).

[2] *J. Elisha Mitchell Sci. Soc.* **56**, 241–82 (1940).

niques, and found that the presumptive differences are reversible
and accommodate themselves to the decrease in size (Fig. 22).
Also we found a similar process for *D. mucoroides* and furthermore
showed that regulation takes place constantly in the normal
development of *D. mucoroides*. Since it forms a stalk continuously,
it is steadily losing cells at the anterior end; yet if measured at any
point during its migration it shows proportionality and falls on

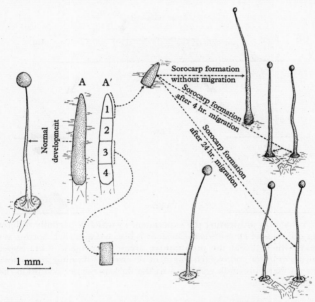

Fig. 21. Comparison of the fruiting of entire cell masses of *Dictyostelium discoideum*
with different fractions of the same. If apical fractions fruit immediately they show
abnormal proportions, but with some migration the normal proportions are
resumed. (From K. B. Raper.)

the curve shown in Fig. 20. By suitable methods[1] we have
obtained migrations producing stalks as long as 22 centimetres;
which means that the majority of the cells of the original mass
have turned into the stalk and as migration proceeds the division
line is constantly moving posteriorly, yet always so that the
presumptive areas show the correct proportionality at any one
moment.

[1] J. T. Bonner and M. Shaw, *J. Cell. Comp. Physiol.* **50**, 145–53 (1957).

This should be convincing evidence that there is a delicately controlled spacing mechanism; and now our task is to discover the mechanism. This is at the moment impossible, and all I can offer is a theory which could explain the facts. It has the merit, at least, of showing what kind of an answer is needed.

To state the problem in simple terms, the cells in the front region of a cell mass 'know' how many cells there are behind;

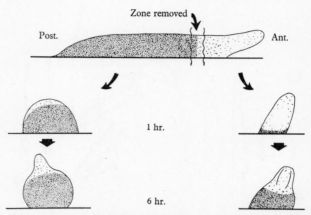

Fig. 22. A diagram illustrating the experiment in which a partially differentiated migrating cell mass of *Dictyostelium discoideum* is bisected and each portion is stained by methods which reveal the presumptive areas, after 1 and 6 hr. respectively. Note that the anterior end of each fragment reversed its staining properties; in one case from the light pre-stalk condition to the dark pre-spore condition, and vice versa in the other.

and if some are cut away, then they can quickly accommodate themselves to the change. The implication is that there is some sort of message that is passed to the front end in a polar fashion and a possibility might be that certain cells, which move more rapidly than the others, are constantly arriving at the anterior end, and the larger the posterior end the more of such cells. Fast-moving cells do exist and can be demonstrated, as well as slow cells that lag behind, so this suggestion is not completely unreasonable. The factor carried by these forward-moving cells might be something as simple as a substrate necessary for a reaction that takes place at the anterior end. The reaction would be the one responsible for the formation of the presumptive stalk condition as well as the stalk itself. This hypothesis is summarized in Fig. 23. Note that there are now three processes which involve a rate: the delivery of the substance (S) at the anterior end (k_1); the formation of stalk (k_2); and the formation of pre-stalk (k_3).

In the case of *D. discoideum* no stalk is formed during migration, therefore k_2 does not exist and

$$k_1 = k_3.$$

This would give (as is the case in Fig. 20) a linear relation between the pre-spore and pre-stalk volume of differently sized cell masses of *D. discoideum*.

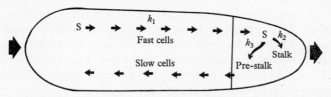

Fig. 23. A diagram illustrating the hypothesis concerning the mechanism of proportional differentiation of cellular slime moulds. S indicates a polar factor which moves forward at rate k_1 and is used to make stalk at rate k_2 and pre-stalk at rate k_3.

In the case of *D. mucoroides* which forms a stalk continuously during migration, the polar factor (S) at the anterior end (k_1) is channelled into two directions in the formation of both stalk (k_2) and pre-stalk (k_3). Therefore there is less S available for pre-stalk and accordingly the pre-stalk of *D. mucoroides* is smaller (as shown in Figs. 19 and 20).

In conclusion it should be noted that the basic elements of this proposed explanation of a controlled spacing mechanism, is the polar movement of cells (the cause of which might be a combination of a gradient of acrasin and the pull tensions within the cell mass) and the presence of a reaction taking place at one end only (the cause of which might be the polarity of the cell movement in combination with some internal gradient system). The fact that some cells are faster or slower than others suggests a variability in the cells of the mass, and it may now be profitable to consider the bearing of this fact upon the evolution of the Acrasiales.

We have already argued that the primitive association of free-living amoebae might have had advantages in masking one another's deficiencies. Concerning the adaptive advantages of the migration process and the final lifting of the spore mass into the air one might make some further speculations. In all cases the morphogenesis spatially separates the feeding, vegetative stage from the fruiting, sporulation stage. Feeding, which involves the phagocytosis of bacteria by thin, delicate amoebae must occur in an aqueous medium. The formation of spores, on the other hand, involves a desiccation of the amoebae which become encapsulated and this conceivably could be more efficiently performed in the atmosphere (although many free-living amoebae form cysts in water). More important is the idea that by lifting the mass away from the region of feeding (and this process is aided by the light and heat tropisms) the spores are in a more favourable position for dispersal. The cellular slime moulds would certainly seem to have a more effective method of spreading than free-living amoebae.

With these basic assumptions as a foundation let us now trace a hypothetical sequence of events from the separate, isolated amoebae to the more complex forms. The first step may have been a sensitivity to certain substances given off by their own kind, which led ultimately to the acrasin-controlled aggregation. Aggregation involves a polar movement of cells and with this polarity comes a gradient: high activity in the front end of the cell mass, decreasing posteriorly.[1]

The next step might have been the production of a pedestal to isolate the spores from the substratum. Relatively solitary amoebae such as *Sappinia* may have individual stalked cells or cysts, but more interesting is the recent discovery of K. B. Raper[2] of his new genus *Acytostelium*. This form has a normal, although not highly organized aggregation, and the resulting small cell mass rises into the air exuding an extremely delicate cellulose cylinder that is devoid of any stalk cells. The point is that all the

[1] This is in line with an argument of Haldane (*Année Biol.* **30**, 89–98, 1954) who, in a discussion of the different kinds of communication between organisms, suggests that the signals between the parts of a metazoan have their origin in the chemical signals of protozoa.

[2] *Mycologia*, **48**, 169–205 (1956).

cells produce stalk material and then subsequently they all produce spores; there is no division of labour.

If now in an *Acytostelium*-like ancestor there is an increase in variability within the cell mass (and the importance of this supposition cannot be overemphasized), then one will expect the beginning of a sorting-out of fast, perhaps high-energy cells and slow, low-energy cells. With this differential ability to move comes a differential ability to produce or contribute in the stalk or in the spore direction.

One has to imagine that there has been some sort of facilitation in this sorting-out, that is, in particular the anterior reaction has become exaggerated, perhaps simply because fuel is now constantly poured on the fire. But the process is so energetic that the fast cells become virtually consumed; they become incapable of further propagation and contribute their hollow frames to give further support to the stalk they have manufactured. At first there is a continuously variable population of fast and slow cells, but by this facilitation connected with the anterior reaction, the fast cells produce a totally new cell type: the pre-stalk and the stalk cell. There the continuous variation results in a discontinuous one, for the labour is finally divided into the two functions: spore formation and the morphogenetic activity of stalk formation.

The situation described thus far applies to *D. mucoroides*. *D. discoideum* has made one further, obvious step forward. For organisms that have ceased all feeding before aggregation, cellulose, which surrounds the stalk, is a precious substance. To migrate 10 centimetres or more is a great expense in both cells and cellulose for *D. mucoroides*, while *D. discoideum* can cover the same distance without forming any stalk. Therefore by changing the timing of the various steps that lead to maturity, it has been possible to migrate greater distances more economically.

In many ways *Polysphondylium* presents the greatest enigma. Again, by an alteration of timing, that is, by delaying the preliminary steps towards maturity until the very last moment, this form has been able to produce a fruiting body with many side branches. But here the adaptive advantage seems especially hard to grasp. Perhaps by breaking up the number of sori into small branches the effectiveness of dispersal is increased, but this is an uncertain hypothesis at best.

The Evolution of Development

In conclusion it might be said that in these proposed evolutionary steps, the most interesting advance is the idea that the cell variability within an organism may increase, but always necessarily in a continuous fashion. But since the variation involves a polar activity (i.e. polar movement) there is an opportunity for reactions to exist at one end of the cell mass that are absent in the other. Therefore, the continuous variation turns into a discontinuous one and the result is a division of labour, a differentiation. Furthermore, the fact that there is a balance between the rates of polar movement and the rates of the anterior reaction means that the spacing of differentiation is controlled and proportionality is the result.

In an abstract sense there is a similarity between this development of slime moulds and embryonic development from a fertilized egg. In both cases there is at the beginning a package of variations: in the slime moulds it is a variable cell population; in an egg it is a variable population of genes and cytoplasmic constituents. In both cases there is the common function or activity of controlled spacing, which involves communication of parts (e.g. polarities, gradients, etc.). In both cases the spacing is achieved in a series of steps, one causing the next. In both cases the spacing and the epigenesis are inherited and passed on from generation to generation. At first it seems remarkable that organisms of such radically different types and origin could have such close parallels in their development. This fact becomes less surprising when one remembers that for each the goal is the same: the production of a large, co-ordinated, and unified adult from a minute reproductive body. The selective forces will always push in the same direction and the inevitable result will be a convergent evolution.

THE EXTENSION OF DEVELOPMENT

In the cellular slime moulds there is first a unicellular stage of separate, independent cells followed by an aggregation of the single cells that co-operate in the development of one unified structure. This unusual life cycle is useful in underlining the fact that the borderline between the development of one organism and the association and interaction of numerous organisms is indeed thin, for the slime moulds would appear to be doing both. If we examine other lower forms we are repeatedly confronted with the problem of individuality, for there appears to be a continuous gradation from single-celled individuals through colonies of varying degrees of integration, and finally to multi-cellular individuals. The problem again arises in animal societies which Emerson[1] calls 'superorganisms', and in plants it arises in the compound filamentous forms such as the fleshy fungi. Development, as we have shown, is in essence the interaction between parts, for it is in this way that the spacing, the patterning of an individual is achieved. But if the individual merges into an association of individuals, then there will be an extension of the principle of development to include these larger associations, for here also the colony or the society is integrated and unified by the interaction between parts. If in evolution there has been an increase in size and an increase in the compounding of living units (into colonies, multicellular organisms and societies) then there must also have been a corresponding extension of development.

There are two main advantages in seeking the function of any living phenomenon. In the first place it isolates that property which is selected for in evolution. Secondly it gives one a common

[1] See chapter in *Structure et physiologie des sociétés animales*. CNRS. Paris (1952).

basis for the comparison of the different mechanisms whereby the function is carried out. We have shown, in the case of development, that its function is the spacing of the constituent parts, so that with an increase in size an organized pattern will appear. The spacing mechanisms transcend the cells and there is a communication of parts. Furthermore the process does not take place in one step but in many. From the point of view of natural selection there is a continuous pressure towards integration along with increase in size and this applies to colonies and societies, as well as individuals. From the point of view of comparative mechanisms, which is of great interest to the embryologist, we are as yet mainly ignorant of the all-important details. But since comparisons themselves are often helpful to stimulate our insights, we will now look into what we know of the mechanisms of integration in associations of individuals, where we can extend the ideas of development.

It is possible to be more specific concerning the adaptive benefits of the association of individuals. In the case of the slime moulds it was pointed out that such aggregative multicellular organisms could have arisen for nutritional reasons. Deficiencies in amoebae might be masked by close association; there is safety in numbers. Numerous other examples could also be used to illustrate the idea that nutritional advantages might be a selective force favouring association. This matter will be considered in detail presently, along with other possibilities, such as mutual protection. Another most important cause of interaction between organisms is sexuality, because fertilization requires methods of bringing the gametes together at the right time and in the right place. Here the immediate advantage to the species is to assure reproduction, while in the case of masking nutritional deficiencies the benefits are purely somatic and help in the growth of the individuals. It would seem then that the interaction between individuals concerns both phases of what we have previously called the 'alternation of generations': the phase of variability production and transmission and the phase of gene action. But of course as far as a species is concerned both the germinal and the somatic generations are part of one scheme and any improvement in the adaptiveness of either phase is an improvement for the whole. Furthermore, the hypothetical proposal for the origin of sexuality in the fungi devised by Beadle and Coonradt (see Fig. 1) suggests that possibly sexuality and nutritional inter-dependence may have arisen simultaneously (in a sort of primitive confusion of the appetites). In the examples given below it will be shown that both the sexual and non-sexual interactions have many aspects in common, all of which are an extension of the method of development within individuals.

A most primitive kind of somatic association occurs in bacteria and fungi where it can be shown that if certain strains exist together, each of which possesses a nutritional deficiency which the other lacks, then both of them can, by the diffusion of these essential substances outside the cell, obtain their full necessary diet on minimal medium, and in this way they both will thrive.[1]

This is essentially the same phenomenon as that discovered by T. T. Puck[2] in his tissue-culture studies of isolated mammalian cells. He found that certain clones derived from single cells would grow only if they were in the presence of other strains or cells. There is a mutual nutritional dependency between the different cell types, and therefore they must all be together for their proper growth and development within the individual.[3]

There is another most interesting case along the same lines from the recent work of P. Brien and M. Reniers-Decoen[4] on the role of the interstitial or *i*-cells of hydroids. They were able to show by some careful experiments that the *i*-cells, contrary to the suppositions of many previous workers, are not directly necessary for growth and differentiation either normally or in regeneration. The *i*-cells are, as was known previously, especially sensitive to X-rays. By using a dose which would kill the *i*-cells but not the whole *Hydra*, normal budding and regeneration after cutting would proceed for a period, and then stop suddenly after some weeks, with the final disintegration of the animal. To prove that this was not a general effect of the radiation on all the cells, they

---

[1] This has been called cross-feeding or syntropy.

[2] Fifteenth Growth Symposium (presented in Providence, R.I., 1956).

[3] Of a slightly different but parallel nature is the evidence from tissue-culture studies that specific cell types must be together in order to obtain inductions and further differentiation. See especially the work of C. Grobstein, Thirteenth Growth Symposium: *Aspects of Synthesis and Growth*, Princeton University Press (1954), pp. 233–56.

[4] *Bull. Biol. France et Belgique*, **89**, 258–325 (1955).

grafted parts of non-irradiated animals (in various ways) on to irradiated ones and were able to keep the irradiated parts from disintegration (Fig. 24). Using histological techniques they demonstrated the loss of all *i*-cells four days following irradiation

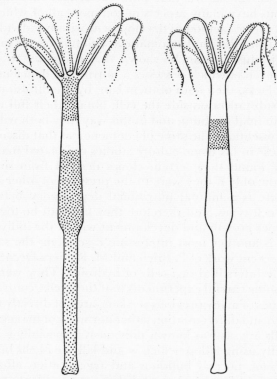

Fig. 24. Diagrams showing the grafting of a ring of irradiated *Hydra* tissue between two pieces of normal tissue (left), and a ring of normal tissue surrounded by irradiated tissue (right). In both cases there was a migration of *i*-cells from the normal parts into the irradiated portion and there was no degeneration of the irradiated tissues. (From P. Brien and M. Reniers-Decoen.)

and furthermore showed that when a non-irradiated portion was grafted on to an irradiated one, there was a migration of the *i*-cells into the irradiated part.

The fact that asexual buds can be initiated or continue to develop in whole irradiated *Hydra,* and the fact that a piece of irradiated *Hydra* will initiate regeneration, are both convincing evidence that these processes can occur in the absence of *i*-cells. It would appear that the function of *i*-cells is rather that of maintenance, or, at least, if it produces substances necessary for morphogenesis these substances can exist in the tissues for weeks before they are depleted. Certainly the idea of many earlier authors that the *i*-cells themselves are necessarily incorporated into the new tissue is wrong, as is the idea that they are essential for induction.[1]

One may speculate further as to whether or not there has been a segregation of cell types which represents a division of labour. The *i*-cell strain and the other strains within the *Hydra* grow side by side, but the other strains are dependent upon some metabolic, synthetic ability which has been retained by the *i*-cells alone. This would be the same phenomenon suggested by the experiments of Puck in mammalian cells where there is also a mutual dependence between cell types. If this interpretation is correct, both examples are instances of cell variability appearing in development, and we assume that this variability is of such a specific nature that the cellular interactions of these particular cell types contribute specific steps which lead to development. The only restriction that must be kept in mind is that these cell differences must all be transmitted through one fertilized egg; in this case we are still talking of the development of a single organism.

Chemical or nutritional dependency is often the basis of symbiosis as well. That is, the development of two associated symbionts may be controlled by the nutritional factors they provide for one another. Both partners are dependent upon each other for their individual development; but now in this case there is an innovation, for the combined structure they produce is a double organism which arose by the communication between parts, although in this case the parts are really separate species, separate individuals.

In the classic case of lichens there is an extremely close association between a fungus and an alga. The fungus is the bulkier member of the pair, and in a sense it affords the housing. The

---

[1] P. Tardent, Roux's *Arch. f. Entw. mech.* **146**, 593–649 (1954); J. Moore (Singer), *Quart. Jour. Micr. Sci.* **93**, 269–88.

alga, on the other hand, lies inside the fungal mycelium, and often the hyphal threads will be seen clasping the algal cells like fingers about a flute (Fig. 25). The conditions of the association in these widely divergent groups may be illustrated by their methods of reproduction. If the fungus produces ascospores, which is often the case, upon germination the resultant hyphae will soon die if they do not meet their algal partner. In some forms there are special reproductive bodies called soredia which are small balls of algae surrounded by fungal hyphae.

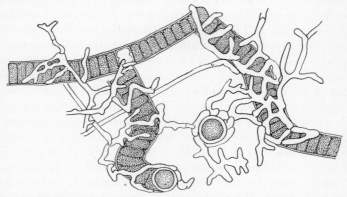

Fig. 25. Drawing showing the association between fungus and alga in two different lichens. (Redrawn from E. Bornet.)

The problem of what each partner provides for the other and whether or not it really is a case of mutual benefit is a more complicated matter. There are those who hold that this is a case of symbiosis, while others argue that the fungus is a parasite of the alga, and still others hold the reverse. The alga is believed to contribute the products of photosynthesis and the fungus is believed to fix nitrogen in peptones from ammonia, and the choice from the three possibilities given above depends upon the relative weight attributed to these two synthetic processes. It may well be that some lichens are symbionts while others are of a more parasitic nature, and as Caullery[1] points out, the spectrum of conditions between true symbiosis and true parasitism is a continuous one.

[1] *Parasitism and Symbiosis.* Sidgwick and Jackson, London (1952).

In any event, nutrition is obviously of prime importance, and this is generally thought to be a usual basis for symbiosis and parasitism in general. There are innumerable cases involving plant parasites or symbionts in animals, plants in plants, animals in plants, and animals in animals. It is especially surprising in the case of symbiosis how little is known of the details of the nutritional requirements, although some cases are being elucidated and much work is being expended in this direction. The nutritional requirements of parasites are far better understood, since this is a matter of greater economic importance as well as a much easier subject to study, for it is entirely one-sided and involves the nutrition of the parasite only.

In these studies it is sometimes found that the requirements of the parasites are very general, but more often there is at least one factor which is specific and brings the parasite and the host together. It occurs in the form of a specific substance given off by the host to which the parasite is in some way sensitive or there is a specific chemical combination between host and parasite. This is not surprising, because host specificity is something that was always recognized as a character of parasitism, and now we have a chemical basis for this specificity. No doubt such specificities have their adaptive value and have been developed by selection. The pertinent point is that we again have an example of specificity associated with reaction pathways, but here the specific chemical interaction involves different organisms which are in association and develop into a combined individual rather than the combination of parts within the body of one organism.

There are undoubtedly other factors in parasitism and symbiosis besides nutrition, although this would seem to be the most important one. Protection or shelter may be concerned in some cases, and it is conceivable that in the lichens the fungus provides an advantageous shelter and moisture-retaining device for the alga. In fact many small parasites which reside in large hosts would seem to obtain their room as well as their board.

Apart from the problem of possible benefits there is also the question of how the organisms become associated, and, in a number of examples reviewed by Davenport,[1] it can be shown

[1] *Quart. Rev. Biol.* **30**, 29–46 (1955).

that there are specific chemicals concerned. He cites the interesting case described by Keeble and Gamble where the free-living flagellated stages of an alga are chemotactically oriented towards the eggs of a flatworm, into which they then penetrate and reside as symbionts. He himself has made some valuable experimental studies on the specificity of chemical recognition of hosts by marine commensal polynoid annelids. By giving the worms a choice in a Y-tube coming from two aquaria he was able to show that in many cases the host gave off specific substances to which the worm responded.

Thus far all I have done or intend to do here is to point out the fact that there are chemical relations, often nutritional, and often remarkably specific between host and parasite or between two members living in symbiosis, and that these specific factors are responsible for the development of the dual organism. I am fully aware that this is a vast subject which has filled many volumes, but I want only to show briefly that it fits into our argument. The subject is vast because it is such a widespread phenomenon; it is an important kind of interaction between individuals.

Let us now turn to the non-somatic kind, which is equally important but for different reasons. Parasitism and symbiosis have arisen as modes of existence (or often modes of feeding) of phenotypes. There are equally successful non-parasitic or non-symbiotic organisms that exist in the same taxonomic groups in which we find the symbionts and parasites. An organism can live and eat in partnership with another kind of organism or it can live a free and isolated life; parasitism or symbiosis is simply one of two possible ways of existing. On the other hand the type of interaction between individuals involved in the transmission of variation is an absolute requirement of the genetic system. In higher forms it is impossible to have recombination without mating and mating involves the coming together of gametes. Fertilization is then the ultimate result of the sexual interaction between organisms, but there are many facets of this problem which we shall take up in their turn. The whole subject may be characterized as one which, during the course of evolution, has become increasingly dependent on complex interrelations of parts, and presumably the complexity assures a more efficient and dependable mechanism in which the errors diminish and the certainty of accomplishing a particular result increases.

The very act of fertilization itself, which might be considered the primary act in the sexual interaction between individuals, involves in those organisms in which it has been carefully studied, both specificity and a chemical interaction between egg and sperm. Starting with the work of Lillie, and more recently the work of Tyler[1] and many others, it is known that eggs of echinoderms produce a substance called fertilizin while the sperm possesses a complementary antifertilizin. (This antifertilizin is also present in the egg, but the role of the egg antifertilizin in fertilization is not known.) Presumably the sperm adheres to the egg because of the fertilizin-antifertilizin combining reaction. Furthermore there is a correlation between the specificity of this antigen-antibody type of reaction, when the extracts from different species are cross-tested, and the ability of the eggs and sperm of different species to cross-fertilize one another. Cross-fertilization is somewhat more selective however, so it is thought that the chemical specificity of the fertilizin-antifertilizin reaction is an important, but not the sole, factor that prevents cross-fertilization between distantly related species.

Besides the attachment there is also the problem of whether or not the egg attracts the motile sperm by chemotaxis. Originally this was thought to be a widespread phenomenon mainly because of Pfeffer's successful demonstration of the process in the spermatozooids of ferns. He was even able to show chemotaxis towards a solution of malic acid, firmly establishing the chemical basis. There have been similar although not such striking demonstrations in other plants, but in animals the situation is quite the opposite, and Tyler for instance states that 'it has not, as yet, been adequately demonstrated for animal spermatozoa'.[2]

Another important chemical which has been identified in sperm is a lytic agent which has the ability to break down the viscous coating and the membrane of the unfertilized egg. In

---

[1] For a good review of this subject, see Chapter v–1 by A. Tyler in Willier, Weiss and Hamburger, *Analysis of Development*, W. B. Saunders, Philadelphia (1955). See also Lord Rothschild, *Fertilization*, Methuen (1956).

[2] *Analysis of Development*, p. 181.

mammals this substance has been called hyaluronidase, for it has the ability to break down the intercellular cement, hyaluronic acid, and in sea-urchins Runnström has evidence that a fatty acid is the active lytic agent. In the case of mammals there does not appear to be much species specificity of the egg lysin, but in molluscs on the other hand there is evidence for a narrow range of action. In all these cases the obvious interpretation is presumed to be true, namely, that the lysin helps in the penetration of the sperm into the egg.

Once the sperm reaches the egg, the interactions become more complex and difficult to analyse. There is a rapid change in the egg cortex which probably prevents polyspermy. There are in different organisms various types of fertilization cones which seem to be a reaction of the egg to the sperm and a further help for the sperm to become incorporated into the egg. And finally another obvious morphological change is the lifting of the egg or fertilization membrane, but the significance of this is more difficult to grasp since in some species if this membrane elevation is inhibited normal fertilization and development nevertheless ensues. However, as Tyler indicates, some importance should be given to the observations of Hultin that such a treatment makes the egg more susceptible to cross-fertilization, and therefore it may again be involved in the specificity of the fertilization process.

It should be pointed out that there are many aspects of fertilization which we can directly observe, but they are nevertheless not fully understood, and furthermore there are undoubtedly many we cannot see and have not yet detected by an experimental means. Yet despite this, what is known amply fulfils our expectations: namely, there is an interacting system of actions and reactions, and a considerable degree of specificity associated with some of these steps.

If we turn now to mating in lower forms, we find the interesting work of C. B. Metz[1] and others in the ciliates. In *Paramecium* Sonneborn first showed that conjugation was the result of the coming together of individual opposite mating types, and that these types were of a fixed, inherited constitution. The filtrate from one type will have no effect on the individual of the opposite

[1] Chapter IV in A. H. Weinrich, *Sex in Microorganisms*, AAAS. Washington (1954).

type; there is no evidence of any diffusible substance in *Paramecium*. The agents appear to be of a specific antigen-antibody nature that apparently lie on the surface of the cilia. In other words, the primary contact between conjugating cells is effected by the adhesion of the cilia and then subsequently the male nucleus pushes through the cell membranes from one individual to the other. Originally Woodruff and Boell and later Metz showed that if cells of one mating type are killed by appropriate means, it was still possible for the pairing to take place. However, there was of course no transfer of protoplasmic material. Also Metz obtained from Sonneborn a mutant stock of animals which could adhere to its opposite mating type, but was incapable of proceeding further in the conjugation process. There is then in *Paramecium* a high degree of specificity associated with the surface of the cilia, and indications of a stepwise process. Before leaving ciliates, it should be mentioned that Kimball has shown that the filtrate of certain clones of *Euplotes* will induce the animals of certain other clones to conjugate, indicating that ciliates do not all follow the exact pattern of *Paramecium*.

In the lower plants there are a number of examples of far more complex reactions in mating and these have been comprehensively reviewed by J. R. Raper.[1] For consideration here I will choose Raper's own experiments on the water mould *Achlya*.

In this form there is not only a mechanism which brings gametes together, but the whole male and female plants (in heterothallic species) respond to one another in a series of co-ordinated steps that ultimately result in fertilization (Fig. 26). The techniques he used included the placing of one plant in the filtrate of another at different moments in their life history (sometimes by a steady flow of one small aquarium spilling into another), as well as separating the oppositely sexed plants by semipermeable membranes. With many such experiments, carefully devised and planned, he was able to identify a whole series of hormones, the actions of which are summarized in Fig. 27.

First comes the A-complex which is contributed both by the male and female plants; it is in itself a delicate control system in which the female produces substances which initiate the

[1] *Bot. Rev.* **18**, 447–545 (1952).

production of antheridial hyphae in the male (A and A²) and the male produces substances which regulate the magnitude of the effect by having a balance between an inhibitory substance (A³)

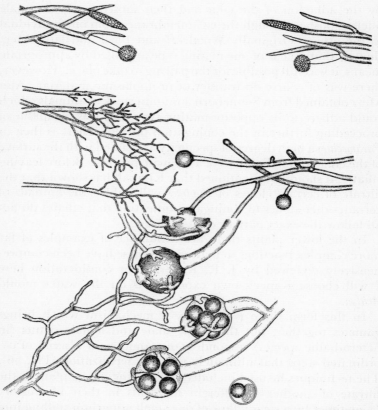

Fig. 26. A semi-diagrammatic drawing of the sequence of events leading to fertilization in a heterothallic species of *Achlya*. The male plant is on the left-hand side and the female on the right. (From J. R. Raper.)

and an augmenting substance (A¹). Once the antheridial hyphae form they in turn produce a hormone (B) which stimulates the formation of oögonial initials which in turn produce a substance (C) that is capable of both orienting the antheridial hyphae by chemotropism and causing the delimitation of the antheridia.

The Extension of Development

Now the latter begin producing a hormone (D) which stimulates the delimitation of the oögonia which is followed by the differen-

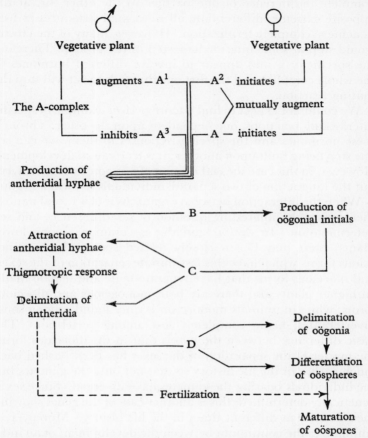

Fig. 27. A scheme showing the steps in the hormone interactions leading to fertilization in a heterothallic species of *Achlya*. (From J. R. Raper.)

tiation of the oöspheres. The plant is now ready for fertilization, which ensues immediately.

Raper[1] has also investigated the specificity involved in such a mechanism as it concerns different related species, and he

[1] Chapter VI in E. G. Butler, *Biological Specificity and Growth*. Princeton University Press (1955).

finds that by crossing different species he can, depending upon the cross, either obtain a block right at the beginning and have complete indifference of one partner to the other, or, at the opposite extreme, differentiate all the sexual structures as well as achieve complete fertilization. However, many of the crosses would stop at some intermediate step in the process. Therefore the specificity would appear to involve different hormones in the whole complex, and the first one that fails to act will stop the mating reaction.

We could not expect to find a more perfect example of specific interlocking reactions, one dependent upon the other. Through these hormones and the specific responses to them, we can see one step being built upon another; it is true causal development. However, in this case the elaboration of a zygote is not involved, but the interaction of two separate individuals to form a zygote.

When the interaction between organisms is of a sexual nature, there arises the interesting question of sex differences and sex determination. In *Achlya* (and this is common among lower plants) there may be, in closely related species, both monoecious forms which have this last-minute separation of the sexes, and dioecious forms that have separate male and female thalli. In higher plants also there are both monoecious and dioecious forms, while in animals monoecism is only found among some invertebrates, being completely lost among vertebrates. The basic difference between the two is that in the dioecious forms the matter of the separation of the sexes has been pushed back much earlier in the life history so that not only the gametes but the individuals bearing the gametes have diverged. Since sex is genetically determined, this is another case of the genes showing their activity at different times in the life history. Moreover, it emphasizes the continuum between the development of an individual (when the sex differences appear in ontogenesis) and the extension of development in the broad sense of interaction between individuals (the steps concerned in the mating reaction).

Regarding the adaptive significance of dioecism, I turn to the well-chosen arguments of Mather,[1] who points out the likelihood that this is a mechanism for ensuring a certain degree of out-

[1] C. D. Darlington and K. Mather, *Genes, Plants and People*. George Allen and Unwin, London (1950).

breeding. This idea is bolstered by the fact that in monoecious organisms there have arisen additional special mechanisms for increasing outbreeding. The pertinent point here is that in both cases there are methods of preventing self-fertilization and close inbreeding and these methods again constitute good examples of interactions between individuals, during their over-all development.

In dioecious fungi there is an elaborate gene-controlled mechanism which prevents certain crosses. The opposite sexes are haploid and in the simplest case there are two allelomorphs for one locus, A_1 and A_2. In the haploid mycelia, A_1 is incapable of nuclear fusion with another A_1 and therefore all zygotes will be A_1A_2. This system is rather rudimentary in that it only prevents self-fertilization of a haploid hypha but puts no restriction on brother-sister fusions. In the higher fungi there are two additional mechanisms each of which preserves the prevention of self-mating, but decreases the likelihood of brother-sister matings. One involves an increase in the number of allelomorphs at one locus; for example, suppose we now have A_1, A_2 and A_3. This will give three types of zygote, A_1A_2, A_1A_3, A_2A_3, and here a haploid hypha has a greater likelihood of fusing with a non-sister pair. If the number of allelomorphs is further increased, the probability of the desirable non-sister pairing also increases.

The other mechanism is an increase in the number of loci. If there are two loci, which is the case in many basidiomycetes, then the haploid hyphae may have the following constitution (provided each locus has only two allelomorphs): (1) A_1B_1, (2) A_2B_2, (3) A_1B_2, (4) A_2B_1. From these gametes there are only two possible crosses, (1) and (2), (3) and (4), because only one zygote is possible $A_1A_2B_1B_2$. However, this mechanism by itself reduces the chances of an outcross as well as a sister cross. But if now there are multiple alleles for the two loci an effective incompatibility system is possible, and such are found in the basidiomycetes.

In angiosperms, where monoecious forms are common, and therefore the danger of inbreeding is great, one finds a number of interesting incompatibility mechanisms. They fall into two major groups: a mechanical group in which the gametes in the flower ripen at different times, and a chemical group where, as in fungi, genes control incompatibility substances which prevent certain crosses. The first category is commonly observed; either the stamens or the pistils will ripen first and so the pollen of an individual flower cannot possibly fertilize its own eggs.

Chemical incompatibility methods depend to a large extent on the tissue of the style, which separates the egg from the pollen, for the latter has to grow through it. As Mather puts it 'the stylar tissue, and perhaps also other somatic tissues of the ovary and ovule, acts as a sieve which stops the tubes of certain genetical types of pollen, while permitting others to grow to successful fertilization'.[1] It is important to note that this stylar tissue is diploid and therefore the incompatibility reaction may be between the haploid sperm and the diploid style, while the egg does not appear to be involved. In other cases it is not the genetic constitution of the pollen that is important, but the constitution of the diploid tissue of the stamen, and therefore compatibility requires the male and female tissues to have the right combination of genetic factors. These factors operate in much the same way as those in fungi which we have discussed, and there are cases with one or two loci and two or more allelomorphs at each locus. The advantages of the various gene systems are the same as in the fungi.

In a recent review J. R. Raper[2] has pointed out that there are a number of different kinds of chemical mechanisms in angiosperm incompatibility. In one kind described by Lewis in *Linum*, osmotic pressure affords the key, and there must be the proper relation between the osmotic pressure of the pollen and that of the style tissue in order to obtain penetration of the pollen tube. Another kind, also demonstrated by Lewis, is found in *Oenothera*, where the pollen-style compatibility is like an antigen-antibody reaction.

All the examples so far of communication or interaction between organisms have been on what we might call a chemical level. This does not mean that all the details of the chemical reactions are known, but that in general there is evidence of stimulating substances and specific responses to these substances. As in the development within an organism, there is, in this extraorganismal development, evidence of actions and reactions

[1] C. D. Darlington and K. Mather, *op. cit.* p. 121.   [2] See p. 81.

occurring in successive steps. Since the responses often bear a specific relation to the stimulus, they must have evolved in close association, and the result is a specialization of the type found in individual development. Also in both there is a pattern; in the extension of development the pattern includes all the individuals in association, whether the association be somatic or sexual.

But all interactions between organisms are not hormonal. As one would expect from a common function, the mechanisms by which it is carried out are varied. We will now consider the most complex and extreme kind of interaction, namely, those promoted by behaviour patterns in animals. This subject has, in the past few years, received a fresh impetus from the work of such men as Lorenz, Tinbergen[1] and a number of others. One of the central themes of what might be termed the new school of animal behaviour is the idea that in nature animals become conditioned, or especially sensitive, to certain stimuli, and that their behaviour is partially a result of instinctive reactions to these stimuli. There are then two elements in this approach: one is that the stimuli and responses are specific and built up to meet a special set of circumstances (they are canalized), and the other is that the reaction to them is not premeditated or thought out in any way but is innate or instinctive. These dual notions are contained in the Lorenz-Tinbergen terminology where the stimulus is called the 'releaser', implying that the stimulus can elicit a specific response, and the response mechanism is called the 'innate releasing mechanism', where the built-in, automatic quality of the response is stressed. A series of noteworthy studies have been made to determine what qualities of the stimulus are especially significant in prodding the innate releasing mechanism, and a few examples of such studies will help to clarify the nature of these stimulus-response mechanisms.[2]

---

[1] See N. Tinbergen, *The Study of Instinct*, Oxford University Press (1951); and *Social Behaviour in Animals*, Methuen, London (1953).

[2] In this discussion the terms 'releaser' and 'innate releasing mechanism' are deliberately simplified to be used synonymously with stimulus and response. This will undoubtedly greatly distress many workers in the field of animal behaviour, but my purpose here is not to present a critical analysis of ethology, but merely to give illustrations of communication between animals that involve behaviour patterns. (See Haldane, *Année Biol.* **30**, 89–98, 1954, for a pertinent discussion of animal communications.)

Tinbergen has made an interesting analysis of the pecking response of herring-gull chicks. Apparently this pattern of behaviour is not learned, but the young chick is born with an innate releasing mechanism. The pecking is directed at the bill of the mother, and its immediate purpose is to beg for food. If cardboard models are substituted for the mother it is also possible to obtain a response, and furthermore by changing the model in different ways one can analyse what property of the model is the

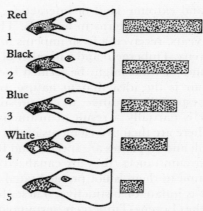

Fig. 28. Models of herring-gull heads with bill patches of various colours (1–4) and without a patch (5). Columns to the right indicate frequencies of begging responses released by the models. (From N. Tinbergen and A. C. Perdeck.)

effective stimulus. There is no need to look at all of Tinbergen's experiments, but one will serve to illustrate the point. There is a red dot on the parent herring-gull's bill, and if dots of different colours are placed on the cardboard model the effectiveness of each can be tested, and as can be seen from Fig. 28 red is the most effective while black, blue, white and no dot are progressively less so.

Another example might be the study of egg preference in birds, which again has been developed and extended by Tinbergen. Here the interesting fact is that the normal colour and shape of a bird's egg may not be the most stimulating, but it is possible artificially to insert a foreign egg near the nest that is more attractive. This preference may be on the basis of colour or size; for example the oyster-catcher prefers eggs of especially large size, so large that the bird cannot possibly brood over it, and yet it will repeatedly attempt to do so, all the while ignoring its own egg (Fig. 29). It may at first seem odd that a releasing mechanism stimulated by abnormally large eggs could be evolved, but in fact there is probably no danger in such an extraneous stimulus, because it is unlikely that normally a large gull will place its egg near an oyster-catcher's nest. Not all examples of such studies have been made with birds, but similar experiments have been devised for insects and fish.

Fig. 29. An oyster-catcher reacting to a giant egg in preference to a normal egg (foreground) and a herring-gull's egg (left). (From N. Tinbergen.)

Here again with behaviour patterns there are sexual and non-sexual interactions between individuals. Let us first give some further examples of the latter, and specifically consider those cases of associations of different species. In the instances of parasitism and symbiosis that were already touched upon, I mentioned the work of Davenport on the attraction of commensal worms to partners by the recognition of the chemical given off by the host. This is a perfect example of a releaser (the specific chemical) and an innate releasing mechanism (the reaction of the responding worm), and as Davenport himself points out, the ideas of Tinbergen and Lorenz are especially applicable to the study of animal symbiosis. He cites many examples: for instance, the work of J. H. Welsh on a mite that inhabits the gills of a fresh-water mussel. When the mites are separated and washed free of a substance given off by the mussel they are positively phototactic, but in the presence of 'mussel water' they are negative. So here we have a reversal of the direction of an innate releasing mechanism depending on the presence or the absence of a chemical releaser substance, and presumably this dual response is helpful in bringing the mites to the mussel, although Welsh has evidence for an additional chemotactic factor.

The commensal attachment of sea anemones upon the backs of certain hermit crabs is another instance, and here we see the further possibility that there is not just a single action and reaction but a series, one dependent upon another. Davenport says that the needed extensive analysis on this partnership has yet to be made; but from the work of Cowles it is known that when the hermit crab changes shells, it transfers its anemone to the back of its new shell. The presumption that a chain of events is necessary is based on the fact that the anemone must respond to some special releaser of the crab in order to become detached from the old shell, and the crab must in turn be stimulated to perform this act, and by disturbing the crab during its house-moving the chain can be broken and the process will not be completed.

In considering the importance of releasers in symbiosis and animal societies, Tinbergen says: 'just as the functioning of hormones and the nervous system implies not only the sending out of signals but also a specific responsiveness in the reacting organ, so the releaser system involves a specific responsiveness to particular releasers in the reacting individual as well as a specific tendency to send out the signals in the initiator. The releaser system ties individuals into units of a super-individual order and renders them higher units subject to a natural selection.[1]

These associations of individuals which are co-ordinated by specific releasers frequently operate between members of one species as we have already seen in the case of the herring gull and her chicks. Various phenomena associated with schooling or flocking also illustrate this point; for instance, Tinbergen shows how the presence of a peregrine falcon will bring together

[1] Quoted by Davenport, *Quart. Rev. Biol.* **30**, 29–46 (1955).

the members of a flock of starlings into a tight bunch (Fig. 30). The falcon is the releaser and the response is the coming close together, and this response also is connected with eyesight and a careful control of wing movements so that the starlings stay the proper distance from one another. This co-ordinated flight, which allows the birds to bank and turn in unison, is presumably entirely dependent on each being released by the movements of the others, and it is assumed that, like a miraculous flash, when one shows intentions of moving they all move. In this case the communication between individuals gives rise to a very obvious sort of pattern; it is a forceful demonstration of a spacing mechanism.

Fig. 30. The presence of a peregrine falcon will cause a loosely knit flock of starlings to form a compact bunch. (From N. Tinbergen.)

Fish schooling, which has been studied by Parr and others, operates on much the same principle, and again it is dependent on eyesight for the discrimination of the stimuli. Blinded fish cannot school nor can fish school in the dark.

Non-sexual associations in one species may become far more elaborate in truly social animals. This is a vast and fascinating subject which unfortunately cannot be discussed here in any detail, and all I shall do is to mention a few points to show how it applies to our argument. Some of the most elaborate animal societies are found among insects, especially ants, bees and termites, and in these societies, it is striking to what extent the whole community is organized, integrated, and tightly knit. It is obvious that this integration can only be achieved by the awareness of one individual insect of the activities of others; there is, in these societies, detailed communication between the

[1] See the review of J. E. Morrow, Jr., *Quart. Rev. Biol.* **23**, 27–38 (1948).

individuals and such communication is by stimulus response, or by releaser and innate releasing mechanisms.

Sometimes there is evidence that the stimulus-response mechanism is on a rather rudimentary chemical level, and this is true of the social hormones, now believed to be firmly established in the caste determination of termites. The earlier work of Light[1] and others and the recent work of Lüscher[2] demonstrate that, for instance, a king and a queen of a colony produce substances which inhibit the workers from developing into secondary reproductives. If the king and queen are eliminated, the inhibitory substances are lost, reproductives develop which then in turn inhibit further transformation of workers into reproductives. There are additional interesting facets of the phenomenon, and of course no importance should be attached to the fact that it is an inhibition rather than a stimulation, for the two are one and the same in general terms of chemical control. The passage of the chemicals between individuals is presumably by the constant licking of one termite by another, and this licking mechanism involves releaser-controlled behaviour patterns. But the action of the social hormone itself is probably more directly upon the chemical events which lead to morphological changes during moulting.

In the work of Schneirla[3] upon army ants there are a number of good examples where the releaser has been isolated, and recognized. For instance Schneirla has been able to find the factors which determine the changes from the nomadic phase to the stationary phase and then the change back to the nomadic phase. It depends entirely upon the development of the brood: when the larvae form cocoons this releases a response to remain in one spot, and when the callows emerge from the cocoons, the workers become frenzied with excitement and respond by extensive wandering. Schneirla has many other examples of interactions, such as the relation of the queen to the workers and the method of column formation and of hunting, all of which fit in with the kind of stimulus-response pattern that we are discussing.

[1] *Quart. Rev. Biol.* **17**, 312–26; **18**, 46–63 (1942–3).
[2] *Sci. American,* **188** (May), 74–8 (1953).
[3] For complete references see T. C. Schneirla and R. F. Brown, *Bull. Amer. Mus. Nat. Hist.* **95**, 267–353 (1950).

The intricacies of these patterns are perhaps best exemplified in von Frisch's[1] great work on bees. The bee has a delicate response mechanism set up for reactions not only to colour in flowers (and to hives) but also to scent. Most remarkable of all are the methods of communication among the bees to indicate the direction and the distance of the flowers. These instinctive dances which the returning bee performs in the hive are highly specific releasers and the response to them is sufficiently selective so that the other workers will stream out and go directly to the new source of nectar. Every detail of this fascinating story points to elaborate stimuli and responses, and elaborate interlocking of these action-reaction systems.[2]

Of course this kind of a complex, releaser-dependent behaviour mechanism is not confined to social insects, but the higher social animals show it as well, although in mammals the plasticity of the behaviour pattern might tend to obscure the simplicity of the stimulus-response mechanism. The point is that as one goes up the scale, instinct does not disappear but becomes obscured by thinking, by problem-solving. This increases the possible number of types of response to a particular situation, and in an intelligent animal the fact that by thinking he can choose from alternative courses of action is obviously of high adaptive value. By this increased number of degrees of freedom in the response, the chances of blindly falling into a trap may be reduced, and none of us doubts the selective advantage of intelligence. Instead of increasing complexity by increasing the number of interlocking, rigid, causal stimuli and responses, in intelligent animals the complexity is increased by elaborating the response mechanism. Perhaps one of the most remarkable features of the brain and its involved workings is that its pattern has itself undergone a development, for all its details are transmitted through the genes, through the single-celled gametes, from one generation to the next.

[1] *The Dancing Bees.* Harcourt, Brace, New York (1955).

[2] It should be pointed out in this case (and the principle applies in varying degrees to the many examples presented here) that there is, as von Frisch has shown, a great deal of learning along with the innate reactions. For instance, there is an innate response to scent, but a specific scent of a particular flower is learned.

Let us turn now to the association between animals which is strictly sexual, that is, connected with the specific problem of bringing the sexes together for the fusion of the gametes. We have already discussed this association at the level of the gametes in fertilization and at the chemical level in plants and lower animals where the coming together was aided by interdependent responses to diffusible chemical stimuli, and where special specificity mechanisms were devised in the form of incompatibility factors to prevent excessive inbreeding. In higher animals the problem of bringing just the right individuals together at just the right time is no less crucial, but here again the interlocking mechanisms have become behavioural as well as chemical, and some of the finest examples of releaser mechanisms are found in courtship.

Unless egg and sperm can be ripe and in close proximity at all times, which would be a wasteful and unlikely situation, it is important to have a timing mechanism by which they can be brought together by both sexes when they are both ripe. In lower animals such mechanisms may depend solely on external stimuli; for instance the hydroid *Hydractinea* will shed its gametes after being in the light for 55 minutes (provided this was preceded by a number of hours of darkness). In birds and mammals light also plays an important role; it is now known that the relative lengths of day and night, which change with the time of year, stimulate the animals into their sexual period.[1] In this way these forms which have reproductive seasons are conditioned in this essentially automatic fashion. But this is not all that is necessary for timing; it is only a beginning, for then there must be a method of bringing the sexes together and having them shed the gametes simultaneously or induce one another to copulate at the same time.

A good example of such courtship, extensively studied by Tinbergen, is that of the three-spined stickleback. This small fresh-water fish also has a definite sexual season. As this period approaches the males turn a deep red, separate from the other fish, select a territory, and begin to build small tent-like nests of

---

[1] This is also true of flowering in higher plants. It is a fairly universal phenomenon and, judging from the spectral studies of the light effects, the photochemical mechanism may be similar for all organisms.

hollows can do so safely. At first the chicks of a robin or a thrush will gape when the nest is jarred, but once their eyes are open they will respond to various types of visual cues. The gaping in turn stimulates the mother to cram some food down their mouths, and the effectiveness of this releaser is seen in the case of the European cuckoo. This parasitic bird places its egg in the nest of some other, usually smaller, species, and the cuckoo chick has the ability to give an especially conspicuous and flamboyant gape which is so effective in eliciting the feeding response in the mother that her own chicks will starve. This is a case of parasitism by an innate releasing mechanism.

The relationship between parent and offspring illustrates another important fact, and that is that the sensitivity of the parent to its offspring is not constant throughout the years, but is conditioned by internal hormone changes so that it is sensitive only during the critical periods of rearing the young. In fact these changes are so delimited in time that a gull will only brood eggs after mating, and only feed chicks after a period of brooding, and if birds that have just started to brood are given chicks they will ignore them. This internal conditioning is a stepwise process which dovetails with the external events.

In the case of the growing young there are also internal changes which alter its relations to the external world and its association with its parents. There is in effect a development of behaviour patterns, or a development of the ability to respond to stimuli. These abilities do not arise full blown in the egg but appear somewhere on the road to maturity; they are an integral part of the elaboration of the phenotype. The nervous system and the brain develop, and along with this development comes the progressive appearance of innate releasing mechanisms.

There is one interesting parallel here between the development of behaviour patterns and the physical development of the soma. The egg of the brown alga *Fucus*, when shed, is devoid of any polarity; the polarity may be imposed subsequently by various agents such as light and heat.[1] But these agents can act only during a certain critical period, a number of hours after fertilization; for if the stimulus is given before this period it has no effect, and if it is given too late, the polarity is already established.

---

[1] For a review see R. Bloch, *Bot. Rev.* **9**, 261–310 (1943).

The Evolution of Development

In the case of animals Lorenz found that there was an analogous critical period in a young goose, for the gosling will follow and become attached to any moving object which confronts it during that period. Ordinarily it would be the mother, but if the mother is removed it is possible to have the bird fix on a bird of another species or the boot of the experimenter. Once this object is 'imprinted', as Lorenz calls it, it cannot be effaced after the critical period and such a bird will show no interest in its own parent. In both these cases there appears to be a temporary plasticity during development in which the growing organism can become conditioned or oriented to its immediate environment. It is readily conceivable that such plasticity has adaptive value in some cases, for in the *Fucus* the rhizoid is sent to the most favourable dark crevice; however, in the goose it is harder to see the importance of having a critical period, although obviously it is of value for the gosling to follow its mother.

My purpose in citing so many examples of interconnexions between individuals is to explore the many implications of the idea that with the association of individuals into groups of various sorts including colonies and societies there has been an extension of the principles of individual development to include the development of the whole association. As I have repeatedly emphasized, the function of development is common to all these larger groupings, the function being the controlled spacing through the communication of parts. In all cases the communication has been elaborated in causal steps in which one step leads to the next with considerable specificity. In individual development the communication was achieved through polarities, gradients and inductions; in compound development the most obvious methods are inductions (or chemical stimuli in general) and behavioural patterns dependent upon the central nervous system.

If I were a cultural anthropologist, I might further extend the case to human societies which are achieved by the less automatic, less innate brain of man. Here for the first time in the progressive evolution of more complex development there has been a radical novelty. The fact that human beings can pass information through language and are capable of extensive learning means that the customs, the cultural traditions are not handed down through the genes but are learnt from father to son. Therefore the transmission of ideas, unlike the transmission of all other living qualities, does not have to pass through a small reproductive body in an 'alternation of generations', but can be passed in its full-blown maturity. This means that cultural changes may be rapid, natural selection ineffective, and instead of slow evolutionary progress there is the rapid progress of history. Moreover, this means that in our strict sense, human societies have, by a successful and artful dodge, lost the power of development.

INDEX

Acetabularia, 22
Achlya, 79 ff., 93
acrasin, 28, 43, 65
Acytostelium, 66 ff.
Agaricus, 53 ff.
angiosperms, incompatibility mechanisms, 83, 84
Aronson, L. R., 93
Ascaris, 39
Aspergillus, 55
auxin, 42, 52

Beadle, G. W., 8, 29, 71
Beale, G. H., 16
birds, gaping for food, 96
Bloch, R., 97
Blum, H. F., 3, 4
Boell, E. J., 79
Bonner, J. T., 18, 50, 54, 63
Boveri, T., 39
Brien, P., 71, 72
Briggs, R., 39
Buller, A. R. H., 53, 55

canalization, 44, 85, 95
Caullery, M., 74
Chatton, E., 18, 46–48
Child, C. M., 40, 41, 58
ciliates, 15 ff., 45 ff., 78 ff.
Clark, E., 93
contact guidance, 43
Coonradt, V. L., 8, 29, 71
Coprinus, 53
coremia, 55, 56
Costello, D. R., 57
Cowles, R. P., 88
cross-feeding, 71
cuckoo, 97
cytoplasmic inheritance, 5 ff., 37, 57

Darlington, C. D., 8
Davenport, D., 74, 75, 87, 88
death, 12

de Bary, A., 54
Dictyostelium, 28 ff., 59 ff.
Dobell, C., 26
Driesch, H., 59
Drosophila, courtship in, 93

Emerson, A. E., 69
Ephrussi, B., 5
epigenetic landscape, 44
Euplotes, 18, 45, 49, 79

Fauré-Fremiet, E., 46–48
ferns, chemotaxis in, 77
fertilization, 76 ff.
Filosa, M. F., 32
fish schooling, 89
Foraminifera, 14, 28
Ford, E. B., 38
frog, nuclear transplantation in, 39
Fucus, polarity in, 97
fungi, hyphal fusions in, 24

Gamble, F. W., 74
gastrulation, 43
gene action, 36 ff.
Goldschmidt, R., 36, 37
Gordon, M., 93
gradients, 40 ff., 58
Graphium, 56
Green, P., 52
Guilcher, Y., 21

Haldane, J. B. S., 38, 66, 85
Halteria, 48
Hämmerling, J., 22
Harvey, E. B., 4
hermit crab, reaction to sea anemone, 88
herring gull, begging response, 86; brooding, 97
heterocaryosis, 10, 12, 32
heterocytosis, 32
Holtfreter, J., 43
Hultin, T., 78

Index

Huxley, J. S., 1, 11, 38, 62
Hydra, role of *i*-cells in, 71 ff.
Hydractinea, shedding of gametes, 93
Hydrodictyon, 14, 25 ff.

i-cells, 71 ff.
imprinting, 98
induction, 42; in fungi, 24, 55
insect societies, 89 ff.

Kane, K. K., 54
Keeble, F., 76
Kimball, R. F., 79
kinetosomes, 46 ff.
King, T., 39

Levey, R., 54
Lewis, D., 84
lichens, 73 ff.
Light, S. F., 90
Lillie, F. R., 77
Linum, incompatability mechanisms in 84
Lorenz, K., 85 ff., 98
Lüscher, M., 90
Lwoff, A., 21, 46, 47, 48

Mather, K., 82, 84
Metz, C. B., 46, 48, 78, 79
Mitchell, H. K., 37
Moner, J. G., 24
Moore (Singer), J., 73
morphogenetic movements, in animals, 26; in mushrooms, 55; in plants, 24 ff.; their relation to polarity, 42
Morrow, J. E., Jr., 89
mosaic development, 39, 57, 58
mushrooms, development of, 52 ff.; incompatibility mechanisms, 83

Neurospora, 36
Nitella, 52
Niu, N. C., 43

Oenothera, incompatability mechanisms in, 84
oöplasmic segregation, 57
origin of life, 3
oyster-catcher, egg preference, 87

Pandorina, 26

Paramecium, 16 ff., 45; autogamy, 17; kappa particles, 6; mating reaction, 78; nuclear activity, 16; surface structures, 46, 48
Parr, A. E., 89
Pediastrum, 14, 24 ff.
Pfeffer, W. F. P., 77
Pitelka, D. R., 46
plant filament, 21 ff.
Plasmodiophorales, 14
polarity, 42; in ciliates, 46 ff.; in plants, 50
Polysphondylium, 59 ff.
Pontecorvo, G., 12
Puck, T. T., 71

Radiolaria, 14, 28
Raper, J. R., 79 ff., 84
Raper, K. B., 62, 63, 66
Rashevsky, N., 41
regulative development, 39, 40, 57, 58
Reniers-Decoen, M., 71, 72
Rhizopus, 55, 56
Rothschild, Lord, 77
rotifers, 11
Runnström, J., 78

Sachs, J., 11
Sappinia, 66
Schneirla, T. C., 90
sea anemone, reaction to hermit crab, 88
Séguéla, J., 18
selection, 3
sex, evolution of, 8 ff., 76 ff.
Shaw, M., 63
slime moulds (Acrasiales), 15, 28 ff., 43, 59 ff.
slime moulds (Myxomycetes), 14, 28
Sonneborn, T. M., 6, 8, 16 ff., 50, 78, 97
Spieth, H. T., 93
Stadler, D. R., 56
starlings, flocking reaction, 89
Stebbins, G. L., Jr., 11
Stentor, 45, 50, 51
stickleback, courtship in, 93 ff.; territorial fighting, 95 ff.
Stysanus, 56
Suctoria, 19 ff.
surface coat, 43
Sussman, M., 29

102

Index

Tardent, P., 73
Tartar, V., 47, 50, 51
Tinbergen, N., 85 ff., 92 ff.
Tokophrya, 20
Trichurus, 56
Triturus, melanophore migration in, 43
Turing, A. M., 41
Twitty, V. C., 43
Tyler, A., 77, 78

Volvocales, 14, 26
von Frisch, K., 91

Waddington, C. H., 35, 39, 44
Wagner, R. P., 37
Weismann, A., 12
Weiss, P., 43
Weisz, P. B., 45, 47, 49, 50
Welsh, J. H., 87
Westfall, J. A., 46
Wilson, E. B., 39
Woodruff, L. L., 79

yeast, 'petite', mutant, 5